RAMAN SPECTROSCOPY FOR CATALYSIS

RAMAN SPECTROSCOPY FOR CATALYSIS

John M. Stencel
University of Kentucky

Van Nostrand Reinhold Catalysis Series

VNR VAN NOSTRAND REINHOLD
New York

Library of Congress Catalog Card Number 89-5630
ISBN 0-442-20514-7

Printed in the United States of America

Van Nostrand Reinhold
115 Fifth Avenue
New York, New York 10003

Van Nostrand Reinhold International Company Limited
11 New Fetter Lane
London EC4P 4EE, England

Van Nostrand Reinhold
480 La Trobe Street
Melbourne, Victoria 3000, Australia

Nelson Canada
1120 Birchmount Road
Scarborough, Ontario M1K 5G4, Canada

16 15 14 13 12 11 10 9 8 7 6 5 4 3 2

Library of Congress Cataloging-in-Publication Data

Stencel, John M.
Raman spectroscopy for catalysis/John M. Stencel.
p. cm.—(Van Nostrand Reinhold catalysis series)
Includes bibliographies and index.
ISBN 0-442-20514-7
1. Raman spectroscopy. 2. Catalysis. I. Title. II. Series.
QD96.R34S74 1989
541.3'95—dc19 89-5630
CIP

VAN NOSTRAND REINHOLD CATALYSIS SERIES

Burtron Davis, Series Editor

Metal-Support Interactions in Catalysis, Sintering, and Redispersion, edited by Scott A. Stevenson, R.T.K. Baker, J.A. Dumesic, and Eli Ruckenstein

Molecular Sieves: Principles of Synthesis and Identification, R. Szostak

Raman Spectroscopy for Catalysis, John M. Stencel

Series Introduction

Frequently we fail to realize how rapidly science develops, and this is especially true during the past century. About one hundred years ago few chemists were engaged in collecting data to find out what kind of relation there was between physical properties and chemical constitution (S. Smile, *The Relations between Chemical Constitution and Some Physical Properties,* London, 1910; S. Nunziante Cesaro and E. Torracca, *Ambix, 35,* 39 (1988). Without the ability to understand these elementary relationships, there was little hope of understanding such a complex process as that represented by catalysis.

In 1881, the first photographic infrared spectra of organic compounds were recorded by Lt. Abney and Capt. Festing (Abney and Festing, *Phil. Trans., 172,* 887 (1881)). By 1905, Coblentz (W. Coblentz, *Investigation of Infrared Spectra,* Carnegie Institution of Washington, Publication Number 35, 1905) could summarize his extensive investigations of the infrared spectra of some 150 organic compounds. His work, combined with the few ideas that were published earlier, clearly showed that the different arrangement of atoms is responsible for the different infrared curves for various compounds and he therefore concluded that structural differences in molecules could be detected by this experimental technique. However, it was not until the mid-1950s that Eischens and coworkers (R. P. Eischens and W. A. Pliskin, *Advan. Catal., 10,* 2 (1958)) showed that infrared spectra could be easily obtained for the adsorbed species that resulted from the chemisorption of molecules on metallic, oxidic or acidic surfaces. The significant discoveries provided by the application of this technique were summarized in two extensive volumes (L. H. Little, *Infrared Spectra of Adsorbed Species,* Academic Press, 1966; A. V. Kiselev and V. I. Lygin, *Infrared Spectra of Surface Compounds,* John Wiley & Sons, New York, 1975 (Russian Edition, 1972)). A volume, now in preparation for this series, will summarize the more recent advances, with emphasis on the results from studies employing dynamic techniques.

Sir Chandrasekhara Raman received the Nobel Prize in 1930 for the effect that today bears his name (C. V. Raman, *Indian J. Phys.*, ii, 387 (1928)). This technique is especially valuable since it complements the infrared technique and, since those vibrational bands that are infrared inactive are Raman active, provides the possibility to obtain additional information about adsorbed species and the catalytic material. The Raman technique found applications in the years following its discovery, but it was applied to the study of adsorbed compounds only after the discovery and wide application of lasers. The use of lasers made it possible to obtain the needed sensitivity to observe the chemisorbed species. Lasers also provided the impetus to study the Raman spectra of catalytic materials as well. The advances during the past 15 years that are reviewed in this volume represent an important body of knowledge for the catalysis community.

Infrared spectroscopy, in addition to providing catalysis scientists a technique important in its own right, has served to introduce the catalysis community to the value of the characterization of the surface species and the catalytic material. It marked the beginning of a revolution that has led today to the application of more than 50 surface science techniques to the study of catalysis and catlaytic materials. This is clearly the most important aspect of catalysis science during the past two decades. This volume is the first of several volumes that will cover this revolution.

BURTRON H. DAVIS

Preface

The acquisition of Raman spectroscopic information from any type of sample was a slow and tedious process until the introduction of stable and brilliant laser excitation sources in the early 1970s. Since that time, and with the development of holographic gratings and specialized electronic equipment, laser Raman spectroscopy has matured to a flexible, analytical technique that is recognized for its ability to provide quantitative data under a variety of difficult experimental conditions or requirements. One of the most difficult applications for which Raman spectroscopy has been successful is catalysis.

The vibrational spectroscopic cousin of Raman spectroscopy is infrared spectroscopy. The application of infrared spectroscopy to catalysis and the identification of adsorbed species was described in a notable book by L.H. Little in 1966 and in other books since that time. Although journal review articles and chapters in books have been published on the application of Raman spectroscopy to catalysis, no book has been published which is dedicated to the subject. *Raman Spectroscopy for Catalysis* is meant to fill this void and to provide a single reference source which gathers Raman data on catalysts acquired from 1975 to mid-1988.

Included in this book is background information on theory and the equipment needed by catalysis researchers who have not been trained in Raman spectroscopy. The application of normal Raman spectroscopic techniques, and the application of more exotic and still developing resonance, time-gated, and Fourier transform techniques, are described in which catalyst speciation during synthesis, activation and reactor activity testing has been delineated. These data form a basis of knowledge to be applied to the evermore demanding requirements which the catalysis community meets in order to understand its materials and the reaction paths and mechanisms to which such materials are applied.

The educational training in Raman spectroscopy which I received at Fisk University and Vanderbilt University during the mid-1970s introduced me to its vast potential as a bulk analysis technique. Only later during my association with scientists at the Pittsburgh Energy Technology Center did my appreciation of the surface analysis potential of Raman spectroscopy develop. For these prior and continuing industrial and university collaborations I am extremely grateful. Special thanks are also due to Burtron Davis who prompted me to write a book which I felt was needed in the truly multidisciplinary research area of catalysis.

Contents

RAMAN SPECTROSCOPY FOR CATALYSIS

1
Introduction and Background

Raman spectroscopy discovered over 60 years ago by C. V. Raman (1), who was awarded the Nobel prize in physics for his work in 1938—has undergone a renaissance within the last 15 years, with the development of novel instrumentation and theory. This renaissance has included its application to the field of catalysis, where, in 1977 and 1978, the research team of Brown, Makovsky, and Rhee (2–4) published the first successful observation of Raman spectra from a highly dispersed surface layer on an amorphous Al_2O_3 support. Almost simultaneously with their publication, Fleischman, Hendra, and McQuillan (5), in addition to Jeanmaire and Van Duyne (6) and Albrecht and Creighton (7), showed that Raman spectra of molecules adsorbed on various noble metals were easily monitored at submonolayer levels, and that this ease of observation was the result of immense enhancements in scattering cross sections of the adsorbed molecules.

It is somewhat surprising that normal Raman spectroscopic techniques, as performed by Brown, Makovsky, and Rhee and even by Fleischman, Hendra, and McQuillan, were not utilized earlier in catalysis research, for example, nearer the time when lasers were introduced as Raman spectral sources around 1970. It is possible that Raman spectroscopy took a back seat to its long-standing cousin, infrared spectroscopy. As early as 1954 Eischens, Francis, and Pliskin (8) had shown the usefulness of infrared spectroscopy as a tool in the analysis of catalysts and adsorbed molecules. Of course, Raman spectroscopy is complementary to infrared spectroscopy, providing information on molecular vibrational modes that have a changing polarizability instead of a changing dipole moment. The activities of infrared and Raman vibrational modes are not necessarily mutually exclusive, as will be discussed in the following chapters. Indeed, in many cases the type of molecular information obtained through Raman experimentation is the same as that found with infrared spectroscopy.

There are some advantages in using Raman spectroscopy that are not available with infrared spectroscopy. For example, the physical size of the sample necessary to obtain a good-quality Raman spectrum is considerably smaller than that necessary for an acceptable infrared spectrum. This advantage is a consequence of the excitation source for Raman spectroscopy—lasers; no broadly and easily tunable infrared source is yet commercially available. In addition, although recent infrared technological advances have enabled microscopic infrared spectral acquisition, where, for example, lateral diameters as small as 10 μm can be used for obtaining spectra, a severe penalty in intensity is paid when the source or infrared beam is masked or condensed to such small areas. In Raman spectroscopy and when using a visible laser source such as Ar^+ or Ar^+-Kr^+, the spot size on a sample can be condensed to spots as small as 30 nm without any penalty on scattering intensities. Hence, depending on the Raman scattering cross section, extremely small sampling volumes can be used to provide information on molecular vibrational modes from any sample, including catalysts.

The Raman scattering cross sections of samples, except those described in Chapter 5 where some data on catalyst surfaces were obtained by surface enhanced Raman spectroscopy, are very small. A small scattering cross section is especially evident for the catalyst supports typically used for dispersing metals and metal oxides. For example, high-surface-area SiO_2 and Al_2O_3 generally do not produce discrete or intense Raman scattering, whereas absorption of an infrared beam usually is severe for these same supports. Hence, the fundamental, internal vibrational modes of supported metals/metal oxides on SiO_2, Al_2O_3, and other supports can be monitored, assigned to specific molecular symmetry and configurations, and the amount of the supported species quantified by use of Raman spectroscopy. At the same time, infrared spectra of the supported phase would be extremely difficult to obtain because the infrared beam is largely absorbed by the support. Such an advantage sometimes is negated by disadvantages. For example, supports such as SiO_2-Al_2O_3, zeolites, and other oxides often produce intense fluorescence which accompanies visible laser illumination. This fluorescence degrades the signal-to-noise ratio of acquired Raman spectra. However, as described in Chapter 3, the fluorescence, whose origin is not thoroughly understood, can be overcome with specialized instrumental techniques, sample preparation, and data processing techniques.

The detectability of supported catalytic phases and of adsorbed molecules is not as well quantified in Raman spectroscopy as in infrared spectroscopy. This lack of quantitation is in part due to the relatively small amount of research that has been performed using Raman spectroscopic techniques as compared to infrared spectroscopic techniques. In addition, the variability of molecular Raman scattering cross sections is considerably greater than the variability of infrared absorptivity of molecules. A variable cross section imparts a degree of flexibility not generally found in infrared spectroscopy although such variability sometimes implies a need to understand a priori the instrumentation or experimental procedure that will enable acquisition of good-quality Raman spectra. For example, the use of resonance Raman or surface enhanced Raman techniques produces molecular detectabilities that in most cases rival the sensitivity of other spectroscopic techniques. Even with normal Raman spectroscopic techniques the detection of high oxidation state and highly dispersed supported phases—molybdena, tungsta, vanadia, rhenia, and so on—at submonolayer loadings is performed routinely. If three-dimensional, supported crystals are formed with these oxides, their detection and quantitation with normal Raman spectroscopic techniques are easily performed. Such sensitivity is discussed amply in Chapter 4 for the standard supports—Al_2O_3, SiO_2, and so forth; other recent and novel applications of Raman spectroscopy, for example, in the delineation of poisoning mechanisms associated with V and Ni deposition on industrial fluid cracking catalysts, also show the flexibility of its use.

The spectral resolution of vibrational bands from solid and supported compounds is a complex function of the instrument used in the measurements and of the properties of the material under investigation. Although specialized spectrometer design can enable band separations as small as 0.01 cm^{-1} to be measured, rarely it is possible to attain such resolution for solid or liquid samples. In these states, the interaction between adjacent structural units or between solvent and solute through dipole–dipole, quadrupole, and H-bonding effects can broaden vibrational bands in some cases to values greater than 100 cm^{-1}. In infrared spectroscopy the influence of H-bonding and/or H_2O on spectral transmission is particularly severe in the 2800 to 3800 cm^{-1} region. Spectral subtraction methods or other tricks enhancing spectral acquisition are required to acquire good-quality data from these samples. Raman vibrational spectra are affected by these factors, but not as much as infrared spectroscopy is influenced.

The advent of the laser as the excitation source of Raman vibrational activity revitalized the application of Raman techniques and has allowed an increased understanding of a variety of chemical and physical systems. As described in Chapters 4, 5, and 6, this revitalization has occurred in catalysis and has included a broad spectrum of sampling schemes and sample types. Time-gated Raman data acquisition, at periods as small as 10^{-12} second, has enabled the delineation of chemical kinetics for transient species not detectable by transient absorption techniques. Molecules in solution and their interaction with solvents, the genesis of interaction of solutions with catalyst supports during and subsequent to impregnation, zeolite synthesis and deactivation, and elucidation of catalytically reactive surface species all have been successfully studied by use of Raman techniques. Presently, gas phase chemistry is also under investigation with specialized Raman techniques such as coherent antistokes Raman scattering (CARS). All of these applications of Raman scattering phenomena have the potential to be and/or have been performed under conditions similar to those employed in actual reaction environments during chemical processing. In general, ultrahigh vacuum conditions are not necessary when Raman vibrational data are to be obtained. Such a decreased restriction implies that in situ Raman experimentation can be considerably less expensive than, for example, types of electron or ion experimentation, and that the information obtained is of direct implication to real-life processes.

Raman spectroscopy is expected to be increasingly influential in catalysis.

Investigations in both fundamental and applied catalysis can be addressed with its techniques. There is tremendous variability in excitation sources, detectors, ancillary data processing and handling equipment, and sampling cells and geometry; and all the devices are relatively inexpensive and readily available from equipment vendors, or can be specially constructed by experimentalists. These options provide Raman spectroscopy a relative advantage over infrared spectroscopy. In addition, theories necessary for the application of various approaches that are possible with Raman spectroscopy have been examined in the literature and are generally well understood. Hence, future users of Raman spectroscopy will obtain important information on the development of catalytic materials and processes, with significantly higher selectivity and activity than are presently available.

Until now no single book has been dedicated to describing the application of Raman spectroscopy to catalysis, although reviews of such application have been published during the past few years. In the belief that a comprehensive compilation can provide catalytic scientists a valuable resource for future use, showing the dramatic potential of Raman spectroscopy, the following chapters have been organized not only to provide an update of Raman applications to catalysis but also to provide insight into the theory and instrumental techniques typically required for specific catalytic applications.

Chapter 2 presents basic theory as well as resources that will provide a greater understanding of the theory. Instrumental techniques, including the basic requirements for spectral interpretation and specialized cells and data processing equipment, are presented in Chapter 3. Chapter 4 details Raman investigations on supported phases that have been published in the last two decades; these data form a firm basis for further elucidation of structural and chemical factors of the solid catalysts used for heterogeneous catalysis applications. Chapter 5 and 6 give a limited view of the potential of specialized, but routinely applied, Raman techniques to adsorbed species and molecules in solution. Finally, Chapter 7 presents future applications, showing some possibilities for Raman spectroscopy in catalysis with presently developing instrumentation.

REFERENCES

1. Raman, C. V., *Indian J. Phys. 2,* 1 (1928).
2. Brown, F. R., L. E. Makovsky and K. H. Rhee, *J. Catal. 50,* 162 (1977).
3. Brown, F. R., L. E. Makovsky and K. H. Rhee, *J. Catal. 50,* 385 (1977).
4. Brown, F. R., and L. E. Makovsky, *Appl. Spectrosc. 31,* 44 (1977).
5. Fleischman, M., P. J. Hendra, and A. J. McQuillan, *Chem. Phys. Lett. 26,* 123 (1974).
6. Jeanmaire, D. L., and R. P. Van Duyne, *J. Electroanal. Chem. 84,* 1 (1977).
7. Albrecht, M. G., and J. A. Creighton, *J. Am. Chem. Soc. 99,* 5215 (1977).
8. Eischens, R. P., S. A. Francis, and W. Pliskin, *J. Chem. Phys. 22,* 1786 (1954).

2
Theory

INTRODUCTION

The scope of this book makes an in-depth theoretical analysis of Raman scattering impossible. However, salient features of the theory will be given to provide the reader with insight into the physical and chemical bases for Raman scattering phenomena and techniques. Benefits of the use of these techniques will be compared and their potential applications described briefly. Raman spectroscopy has blossomed since the advent of use of the laser, so that there are now numerous recognized scattering phenomena—normal Raman, hyper-Raman, inverse Raman, resonance Raman, coherent anti-stokes Raman, surface enhanced Raman, and so on. These phenomena are natural consequences of a generalized Raman theory to be briefly discussed herein.

Molecular symmetry dictates the number and activity of Raman vibrational modes. Symmetry considerations and the calculation of selection rules as obtained from group theory will be described for normal Raman scattering from bulk materials and adsorbed molecules. Specialized theory for surface enhanced Raman spectroscopy (SERS) from adsorbates will be presented to enable an understanding of some of the physical and chemical requirements for its application to catalysis. This theory is available in much greater detail in the references noted at the end of the chapter (1-17).

THEORETICAL TREATMENT

Classical Theory

An electric dipole moment ($\overline{P}$) is induced within a molecule when an electromagnetic wave having an electric field $\overline{E}$ is impinging upon the molecule. This induced dipole can be represented by a power series:

$$\overline{P} = \overline{P}^1 + \overline{P}^2 + \ldots \tag{2.1}$$

where:

$$\overline{P}^1 = \overline{\alpha}.\overline{E}$$

$$\overline{P}^2 = 1/2\ \overline{\beta}\text{: } \overline{EE}\text{, etc.}$$

In general, the direction of the induced dipole is not the same as the electric field, $\overline{E}$. The term $\overline{\alpha}$ is a tensor and is called the polarizability. Typical orders of magnitude of the components in Equation (2.1) are 10^{-40} C/Vm^2 for the polarizability ($\overline{\alpha}$) and 10^{-50} C/V^{2}m^3 for the hyperpolarizability ($\overline{\beta}$) tensor. Hence, for second order contributions from the hyperpolarizability to affect the induced dipole moment to the degree of first order contributions, the incident electric field of the excitation source has to be approximately 10^{10} greater than available from normal continuous mode, gas phase lasers. In gas phase lasers, such as Ar^+ or Ar^+-Kr^+, normal electric field strengths are about 10^4 to 10^5 V/m; from giant-pulse lasers the electric field can be on the order of 10^{10} to 10^{12} V/m. At such large field strengths, second and third order, and even nuclear, processes can become the predominant mode of excitation.

Equation (2.1) implies that the relationship between the P and E vectors can be placed into the matrix form:

$$\begin{pmatrix} P_x \\ P_y \\ P_z \end{pmatrix} = \begin{pmatrix} \alpha_{xx} & \alpha_{xy} & \alpha_{xz} \\ \alpha_{xy} & \alpha_{yy} & \alpha_{yz} \\ \alpha_{zx} & \alpha_{zy} & \alpha_{zz} \end{pmatrix} \begin{pmatrix} E_x \\ E_y \\ E_z \end{pmatrix} \tag{2.2}$$

The above matrix is symmetric; that is, $\alpha_{xy} = \alpha_{yx}$, $\alpha_{xz} = \alpha_{zx}$, and $\alpha_{yz} = \alpha_{zy}$. Non-zero polarizability components determine the directional relationship between $\overline{P}$ and $\overline{E}$. The status of these components is determined by the symmetry of the molecule under investigation (see below).

Classical electromagnetic theory shows that the frequency response of a harmonically oscillating molecular system upon which is impressed an electromagnetic field can be written in terms of the linear induced electric dipole:

$$\overline{P} = \overline{P}(\nu_0) + \overline{P}(\nu_0 - \nu_k) + \overline{P}(\nu_0 + \nu_k) \tag{2.3}$$

This equation implies that three distinct frequency components are present in this molecular dipole. These components give rise to radiation from the dipole at the excitation frequency ν_0 (Rayleigh scattering) and at ν_k from the excitation frequency (Raman scattering). At $\nu_0 - \nu_k$ the radiation is called stokes Raman scattering, whereas at $\nu_0 + \nu_k$ the radiation is termed anti-stokes Raman scattering. The intensity in the stokes and anti-stokes scattering is determined by the population density of the molecular states.

In general, the molecular polarizability will be a function of the nuclear coordinates of the molecule. In terms of a Taylor series expansion, and considering a relatively small vibrational excursion from an atom's equilibrium position, the polarizability of the xy component can be written as:

$$\alpha_{xy} = (\alpha_{xy})_0 + \sum_k \left(\frac{\partial \alpha_{xy}}{\partial Q_k}\right)_0 Q_k + \frac{1}{2}\sum_{kl}\left(\frac{\partial^2 \alpha_{xy}}{\partial Q_k \partial Q_l}\right) Q_k Q_1 + \ldots \quad (2.4)$$

where associated with vibration k, l,. . . are normal coordinates Q_k, Q_l, . . . that describe the vibration with their derivative taken at the equilibrium position, Q_0. The harmonic approximation considers only first derivatives, whereas anharmonic considerations include higher order terms. Anharmonic considerations will account for the possible generation of overtone and combination bands.

From further considerations of the electrodynamics of an oscillating dipole, the intensity of the Raman lines, I, can be described by:

$$I = K(\nu_0 \pm \nu_k)^4 \alpha'^2 E_0^2 \quad (2.5)$$

where E_0 is the electric field strength of the incident excitation source, α' is the change in the polarizability, and $\nu_0 \pm \nu_k$ is the Raman band frequency. The ν^4 frequency dependence on the intensity of the scattered radiation has strong implications for infrared excitation of Raman spectra as compared to visible excitation. In addition, a necessary condition for Raman activity is non-zero components in α_{ij}, where $ij = xy$, for example, and $\alpha'_{ij} = \left(\frac{\partial \alpha_{ij}}{\alpha Q_k}\right)_0$; hence, a plot of a particular polarizability derivative against the normal coordinate must have a non-zero gradient at the equilibrium position for generation of Raman activity. For example, the polarizability changes associated with the vibrational modes of H_2O are schematically represented in Figure 2.1. All vibrational modes are Raman active because they are accompanied by a change in polarizability. All of these vibrational modes are also infrared active because they are accompanied by a change in dipole moment. In the case of a linear molecule such as CO_2, however, the symmetric stretching mode, ν_1, would not be infrared active because no change in the dipole moment occurs. This fact implies that Raman and infrared spectroscopy are complementary techniques.

Quantum Mechanical Treatment for Normal Raman Spectroscopy

Using the Dirac notation, the requirement of a non-zero vibrational transition moment, P_{ab}, is given by:

$$\overline{P}_{ab} \neq 0 = \langle \psi_a | \overline{P} | \psi_b \rangle \quad (2.6)$$

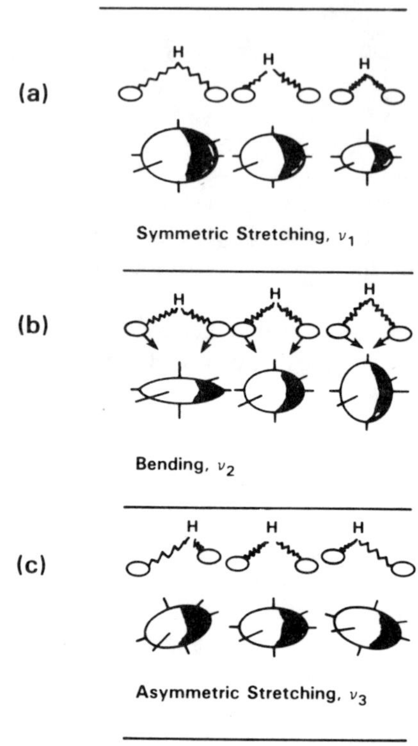

Figure 2.1. Polarizability changes of the vibrational modes of H_2O.

where $\psi_{a,b}$ are final (b) and initial (a) states of the system's vibrational wave function, and $\overline{P}$ is the induced moment operator given in Equation (2.1). Hence, in terms of the xy polarizability component, Equation (2.6) can be written as:

$$(\alpha_{xy})_{ab} = \langle\psi_a|(\alpha_{xy})|\psi_b\rangle \tag{2.7}$$

The coordinates for α_{xy} are fixed to the molecule, and only vibrational states are considered. If so, and with substitution of the Taylor series expansion of the polarizability to include only first order terms as given in Equation (2.4), we obtain:

$$(\alpha_{xy})_{ab} = \alpha_{xy}\ \langle\psi_a|\psi_b\rangle + \sum_k \left(\frac{\partial\alpha_{xy}}{\partial Q_k}\right)_0 \langle\psi_a|Q_k|\psi_b\rangle \tag{2.8}$$

This equation can be used to define whether particular α_{ij} components will be zero. For example, vibrational, harmonic oscillator functions are orthogonal, implying that the first term (relating to Rayleigh scattering) of Equation (2.8) is non-zero only if the quantum numbers do not change; that is, $n_a = n_b$. For the second term, the condition of non-zero values becomes $n_b = n_a \pm 1$, that is, a change in ± 1 of the vibrational quantum numbers.

The extent to which the vibrational states are populated is dependent upon the absolute temperature of the molecular system. Using the Boltzmann distribution law, and assuming that a collection of N molecules is under observation, the ratio of the vibrational intensity in a stokes to an anti-stokes line is given by:

$$\left[\frac{(\nu_0 - \nu_k)^4}{(\nu_0 + \nu_k)^4}\right] e^{(hc\nu_k/K_0T)} \tag{2.9}$$

where ν_0 is the excitation frequency, ν_k is the vibrational frequency of vibration k, K_0 is the Boltzmann constant, and T is the absolute temperature of the sample. This equation is particularly useful when the temperature of the sample is to be measured.

For small molecules, the requirement that the partial derivative of the polarizability at the molecular equilibrium be non-zero is easy to assess. For larger molecules, this requirement is less easy to determine. However, Raman activity also can be assessed by looking at the components of the symmetric tensor (α_{ij}). In terms of Equation (2.3), the condition for Raman activity could be written:

$$(\alpha_{xy})_{n_a n_b} = \langle \psi_b(Q_k) | \alpha_{xy} | \psi_a(Q_k) \rangle \neq 0 \tag{2.10}$$

This requirement implies a non-zero integral if α_{xy} and $\psi_b(Q_k)$ belong to the same symmetry species; that is, under each symmetry operation of the molecule being investigated, the α_{xy} and $\psi_b(Q_k)$ transform in the same manner. Equation (2.10) is a consequence of a harmonic oscillator assumption. However, the consequence of this equation in terms of Raman activity is independent of whether the wave functions are anharmonic or harmonic representations. Group theory must be applied to the molecule for predicting Raman activity.

Time-Dependent, First Order Perturbation Theory

To obtain the electric dipole transition moment associated with the vibrational transition from an initial state a to a final state b, the Hamiltonian of a molecular system can be considered to contain only an electric dipole term and be subject to the perturbation of an electromagnetic field at frequency ν_0. The restriction of using only an electric dipole interaction term does not limit the validity of the solution for the case in which the electric field intensity does not vary over the length of the molecule. Here, in contrast to previous time-independent solutions, the transition moment is found to be a complex number, the real part of which can be written as:

$$(\alpha_{ij})_{ba} = \frac{1}{h}\sum_{r}\left[\frac{(P_j)_{br}(P_i)_{ra}}{\nu_{rb} + \nu_0} + \frac{(P_i)_{br}(P_j)_{ra}}{\nu_{ra} - \nu_0}\right] \tag{2.11}$$

where the summation is over all of the energy states, r, of the system. Because the transition moments in Equation (2.11) always occur as products, there must exist in the system at least one state r that has a non-zero dipole transition moment to both the initial and final states. This statement and Equation (2.11) do not imply that the transition from the initial state a to the final state b occurs in two distinguishable states, that is, via an intermediate state r. Also, Equation (2.11) does not imply the absorption of energy $h\nu_0$ during the transition from $a \rightarrow b$, as there is no requirement to have $h\nu_0 = h\nu_r$. The states r can be higher in energy, lower in energy, or between the states a and b in energy, as is indicated in Figure 2.2 for the Raman transitions in H_2O.

Incorporation of time dependence in the transition moment calculation shows that there will be a phase difference in the scattering between one molecule and the next. Hence, Raman scattering will be spatially and temporally incoherent. On the other hand, no phase difference will exist for Rayleigh scattering; that

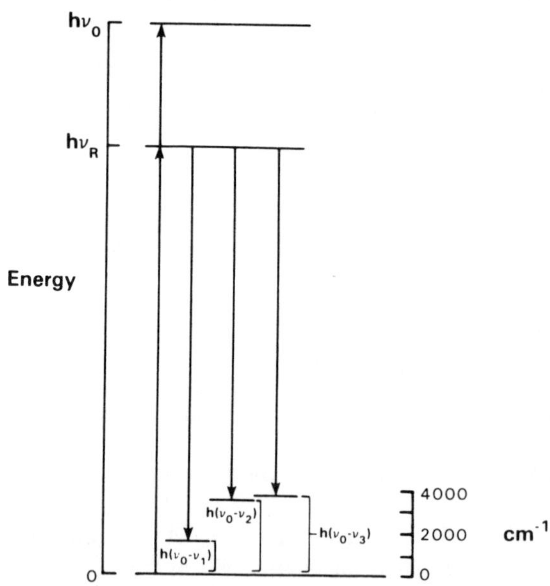

Figure 2.2. Schematic of Raman allowed scattering for H_2O incorporating the intermediate state at $h\nu_R$. The zero energy scale is referenced to the excitation energy $h\nu_0$.

is, it is coherent scattering. In addition, although the formalism of Equation (2.11) involves any energy level r that must have a non-zero dipole transition moment with both initial and final states, the selection rule for Raman vibrational activity can be written in terms not involving the properties of state r. For example, it can be shown that $(\alpha_{xy})_{ba}$ has the same transformation properties as $\langle b|xy|a\rangle$. Hence, the general condition of having a non-zero $(\alpha_{ij})_{ba}$ for Raman activity becomes that the product $\langle\psi_b|ij|\psi_a\rangle$ be non-zero—for example, that $\langle\psi_b|xy|\psi_a\rangle$ belongs to a representation that contains totally symmetric species.

From Equation (2.5), it was observed that the intensity of the scattered radiation was dependent on ν_0^4 or $(\nu_0 - \nu_k)^4$. Using time-dependent perturbation theory, this $(\nu_0 - \nu_k)^4$ dependence still exists as long as ν_0 is small compared to ν_{rb} and ν_{ra}. As $\nu_0 \rightarrow \nu_{ra}$ or $\rightarrow \nu_{rb}$, the intensity of the scattering will depart from the fourth power law. In particular, if $\nu_0 \rightarrow \nu_{ra}$, that is, the incident laser frequency approaches that of an absorption band in the system, the transition polarizability components will have values that are greatly enhanced compared to the situation when ν_0 is not close in value to ν_{ra}. When $\nu_0 = \nu_{ra}$, the anomalously intense scattering is called resonance Raman scattering. By an appropriate choice of excitation frequency, the intensity of the scattered radiation can be enhanced by factors up to 10^5. This implies that Raman activity could be monitored even for molecules of extremely low concentration if a proper choice of excitation frequency were made. However, simultaneous with enhanced scattering cross sections in resonance Raman is the breakdown of normal selection rules; that is, transitions inactive in normal Raman scattering can become active in resonance Raman scattering. This breakdown in selection rules often leads to complications in spectral interpretation that are not experienced in normal Raman scattering.

It must also be mentioned that time-dependent perturbation theory can account for Raman scattering associated with magnetic dipole and electric quadrupole terms in the interaction Hamiltonian. As a result, we can define scattering intensities and polarization from chiral molecules that are different from those from non-chiral molecules. In particular, the difference in scattered intensity due to left or right circularly polarized incident light has been termed circular intensity differential (CID) Raman spectroscopy.

Group Theory

A molecule consists of an assemblage of atoms arranged in a particular geometric configuration. By performing certain geometric (symmetry) operations on the molecular configuration, it is possible to transform the molecule into other configurations that are superimposable on the original configuration (i.e., indistinguishable from it). If these symmetry operations are made and the above transformation occurs, the molecule is said to contain a symmetry element. Five kinds of symmetry elements, as described below, can be defined for an isolated molecule. These can be combined in a number of ways, but the number of resulting sets of all the symmetry elements is limited, and the sets constitute molecular point groups. Only 32 point groups exist, and they describe the symmetry of hundreds of thousands of molecules.

The use of Raman spectroscopy to describe molecular structure in catalysis or other applications largely depends on an understanding of molecular symmetry and basic group theory. Hence, an introduction to these topics will be presented here to give the reader some insight into the material in the following chapters. This presentation is not mathematically sophisticated, but it would be applicable to catalysis research. More rigorous treatment of group theory can be found in many texts (2,13–15).

A symmetry element is a geometrical entity such as a point, line, or plane. A symmetry operation applies the use of a symmetry element to the reorientation of the molecule to a location and/or an orientation indistinguishable from that where it originated. The symmetry elements of isolated molecules include: (1) a rotation axis (C_n), about which the symmetry operation of rotation occurs; (2) a mirror plane (σ), through which a reflection is made; (3) an inversion center (i), which is a reflection through a point on the molecules; (4) an improper axis (S_n), which includes a rotation about an axis plus reflection through a plane; and (5) an identity element (E). For example, Figure 2.3 shows the effects of applying the symmetry elements C_2 (180° rotation), σ_v and σ_v' to the H_2O molecule, on the atoms of which are placed unit vectors in cartesian x,y,z directions. Each of these operations changes to some extent the unit vector directions without chang-

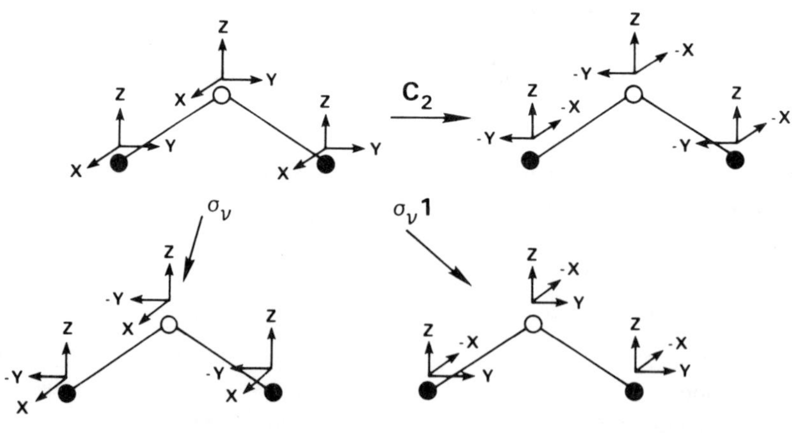

Figure 2.3. Effects of symmetry operations on the unit vectors X, Y, and Z for an H_2O molecule.

Table 2.1. Character table for C_{2v}.

C_{2v}	E	C_2	$\sigma_v(xz)$	$\sigma_v(yz)$		
A_1	1	1	1	1	z	x^2,y^2,z^2
A_2	1	1	-1	-1	r_z	xy
B_1	1	-1	1	-1	x,r_y	xz
B_2	1	-1	-1	1	y,r_x	yz

$$E\text{: } (1)\,(y) = y$$
$$C_2\text{: } (-1)\,(y) = -y$$
$$\sigma_v\text{: } (-1)\,(y) = -y$$
$$\sigma_v\text{: } (1)\,(y) = y$$

ing the location or identity of the individual atoms. Hence, application of the symmetry element has left the molecule indistinguishable from its initial state, except for unit vector directions. The implications of these unit vectors are described in Table 2.1.

An example of an improper axis (S_4) is presented in Figure 2.4 for the tetrahedral species WO_4^{2-}. The S_4 symmetry operation includes a rotation of 90° ($= 360°/n$, where $n = 4$) about the vertical axis followed by reflection through a horizontal plane. The locations of all atoms are left indistinguishable from their initial locations by this symmetry operation. The combination of symmetry elements that can be applied to a molecule is called a point group. In most cases,

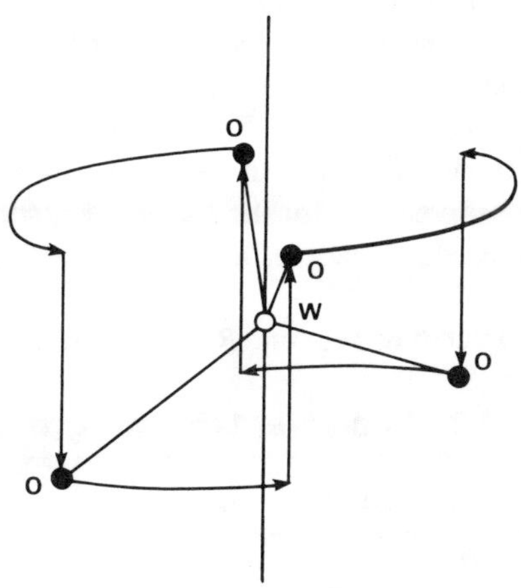

Figure 2.4. Illustration of an improper axis of rotation, S_4, for tetrahedral WO_4^{2-}. This symmetry operation includes a 90° rotation about the S_4 axis followed by a reflection through a horizontal plane containing the W atom.

the symmetry (corresponding to a point group) of isolated molecules, anions, or cations is of primary importance in catalysis.

The system generally used by spectroscopists to label point groups is called the Schoenflies notation, which has an equivalent Hermann-Mouguin notation that is generally used by crystallographers. Hence, a C_{2v} Schoenflies notation, in which the point group contains C_2, σ_v, σ_v', and E elements, is labeled *mm* in the Hermann-Mouguin notation. The number of these elements in a point group is its order; smaller combinations of the elements C_2, σ_v, σ_v', and E also satisfy criteria for a group. This smaller set is called a subgroup, and must contain an integral number of elements that are in the point group. In the case of loss of symmetry of a molecule, for example, by interaction of an adatom with a support, the molecular symmetry descends to that of a subgroup within the point group. Details are presented in many publications on how to define the point group to which a molecule would belong (2,5,6,13–16).

The application of group theory to define the vibrational selection rules of a molecule depends on the use of the irreducible representation of the point group to which the molecule belongs. This irreducible representation will be one of only a few unique multiplication tables of the group. Hence, Table 2.1 shows the complete table for the C_{2v} point group, to which H_2O belongs. The meaning of this table is the following. Consider the effects of the operations on a unit vector in the Y direction, where Z is the direction of the C_2 axis (see Table 2.1). The representation B_2 is no more than an expression of how the unit vector y transforms under the operations of the C_{2v} group. An identical statement can be presented for rotations (r_x, r_y, r_z) about the specified cartesian axes x,y,z and the squares or direct products of x,y, and z.

The character table for a tetrahedral group T_d (as for WO_4, etc.) is given in Table 2.2. This T_d group has five irreducible representations, with the E representation doubly degenerate and the F (more recently defined as T) triply degenerate. Such degeneracy is noted in the next to last and last columns of Table 2.2, where it is observed that for the F_2 representation the x, y, and z unit vectors, and the direct products xy, xz, and y^2, are identical in their transformation.

As described in the preceding section, Raman activity depends on non-zero

Table 2.2. Character table for group T_d.

T_d	E	$6S_4$	$3C_2$	$8C_3$	$6\sigma_d$		
A_1	1	1	1	1	1		$x^2 + y^2 + z^2$
A_2	1	−1	1	1	−1		$(x^2 + y^2 - 2z^2,$
E	2	0	2	−1	0		$x^2 - y^2)$
F_1	3	1	−1	0	−1	(r_x,r_y,r_z)	
F_2	3	−1	−1	0	1	(x,y,z)	(xy,xz,yz)

contributions of the integral of the type: $\langle \psi_b | ij | \psi_a \rangle$, where $ij = xy$, yz, and so on. Hence, the cartesian unit vectors in Table 2.2 suggest that Raman activity could be within the representation A_1, E, and F_2, whereas infrared activity could be within only F_2. The number of possible modes and the active or forbidden nature of these modes have not yet been defined. However, it is generally not difficult to define such activity by use of simple rules on how to manipulate the irreducible representation of a particular system (13,16). Thus, for example, this representation for the C_{2v} group as applied to H_2O is given by $\Gamma_{H_2O} = 2A_1 + B_2$. The A_1 and B_2 representations would be Raman and infrared active, implying the observation of three vibrational modes. The total number of vibrations possible for the three-atom nonlinear H_2O molecule would be $3N - 6 = 3$.

Molecules in a gas are randomly oriented and essentially free of intermolecular or interionic perturbations, but in the condensed state this is no longer true. Here, the orientation of the molecules is such as to minimize the free energy. This leads to an infinite regular repetition in space of identical structural units. These units, through the application of translations, translations-rotations, and/or translations-reflections, can describe the total geometric arrangement of the molecules. In order for these symmetry operations to describe the molecular structure, the environment of a point that is reached after the application of the symmetry operation must look the same in every respect as the environment of the point from which the operation was initiated. The set of all possible symmetry elements that can be applied to a particular ordered molecular system (crystal) is its space group. A total of 230 different space groups are possible, but some are never found in crystals.

The repeat unit of a crystal is called the primitive cell or unit cell. In a crystal, one primitive cell may be carried into an identical cell by a translation. The number of translations would then seem to be infinite because a crystal is composed of many such units. If, however, one considers only one translation and consequently two unit cells, and defines the translation that takes a point in one unit cell to an equivalent point in the other unit cell as the identity, one can define a finite group, which is called the factor group of the crystal's space group. These factor groups are isomorphous with the 32 point groups. The factor group is a supergroup of the groups that represent the symmetry of the environment at atomic or ionic sites. The sets of symmetry elements of a site form the site group; it is a subgroup of the free ion or molecular point group. These sites can be located only on one of the symmetry elements of the factor group, which leaves the site invariant under the symmetry operation, independent of translation.

The whole method of applying group theory to vibrational problems depends upon the fact that normal modes of vibration have special symmetry properties because the wave functions that describe them should remain invariant under the possible symmetry operations. These properties are identical to the properties of certain species of the factor group. Thus, infrared and Raman active modes of a crystal will be determined by the factor group. The correlation of the symmetry species representing the ionic vibrational modes through the site group, to the factor group, limits the number of possible factor groups. For an unknown structure, polarization measurements, the frequency, and the number of the Raman and infrared modes measured can elucidate possible factor groups to which a crystal may belong.

Nonlinear Raman Effects

As indicated earlier in this chapter, the use of pulsed lasers to excite molecular vibrational modes can produce nonlinear scattering. For example, Q-switched ruby ($\lambda = 694.33$ nm) or neodymium glass ($\lambda = 1060$ nm) lasers produce a pulse having a duration of approximately $10–100 \times 10^{-9}$ second. Even though the average power of these lasers would be relatively small, a joule output in 10^{-9} second corresponds to a power of 10^8 watts.

With such high instantaneous powers, the second order induced dipole, $\overline{P^2}$, and the hyperpolarizability, $\overline{\beta}$, become important (see Equation 2.1). The presentation of the generalized theory for nonlinear Raman scattering is beyond the needs of this book, but a simplified classical treatment shows that the result of the influence of $\overline{\beta}$ is to produce scattered radiation at $2\nu_0$, where ν_0 is the laser frequency, and at $2\nu_0 + \nu_k$, where ν_k is the frequency of a vibrational mode. This type of scattering has been called hyper Raman scattering. The general selection rule for generation of hyper Raman activity is that the product of the wave functions for states ψ_a and ψ_b with xyz must contain totally symmetric species. In other words, for vibrational modes involving normal coordinates Q_k, the symmetry class of Q_k and xyz must be the same. Such a selection rule is substantially different from that for normal Raman scattering. Hence, vibrations that are infrared and Raman inactive can be hyper Raman active, whereas all infrared active bands are always hyper Raman active.

Other nonlinear Raman scattering effects include stimulated Raman scattering, inverse Raman, and coherent anti-stokes Raman scattering (CARS). Briefly, stimulated Raman scattering becomes important at a laser irradiance beyond that needed for hyper Raman scattering. Here, a substantial portion (~50%) of the excitation power at frequency ν_0 is converted into radiation at $\nu_0 + n\nu_k$, where $n = 1, 2, 3, \ldots$. Lifetimes of vibrational states can be studied by using stimulated Raman scattering.

The simultaneous illumination of a sample with a high-intensity, pulsed laser of frequency ν_0 and a corresponding high-intensity, broad band source having a range of $\nu_0 \rightarrow \nu_0 + \nu_k$, where ν_k is the frequency of a vibrational mode, can create absorption of radiation from the broad band source at ν_k and concomitant emission at ν_0. This absorption–emission process is called inverse Raman scattering. The laser and broad band source must have spatial and temporal coincidence for inverse Raman spectroscopy. However, the resulting spectra are not restricted to the observance of only one or two excited vibrations, as is the case for stimulated Raman spectroscopy. The transition probability, W, for the inverse Raman effect is given by:

$$W \propto \left[\frac{8\pi h\nu_k^3 n_k^3}{c^3} + \rho_k(\nu_k)\right]\rho(\nu_0) \tag{2.12}$$

where $\rho(\nu_0)$ is the excitation power density, ν_k is the frequency of the Raman photon, n_k is the refractive index at the Raman frequency, and $\rho_k(\nu_k)$ is the Raman power density. Inverse Raman spectroscopy has been applied to the analysis of liquid (18) and gaseous (19) samples. Spectral resolution of an inverse Raman system should be limited only by the line widths of the lasers that are used. In addition, spectra can be recorded in periods as short as 10^{-8} to 10^{-12} second. This latter fact is advantageous in the study of transient species.

Finally, CARS is the result of spatial and temporal coincident mixing of two laser beams that have frequencies of ν_0 and ν_1 such that coherent (laser) radiation is produced at ν_2, where $\nu_2 = \nu_0 + (\nu_0 - \nu_1)$. If ν_0 is varied and the difference $\nu_0 - \nu_1$ matches some molecular vibrational frequency of the system under study, then $\nu_2 = \nu_0 + \nu_k$. Here, the conversion efficiency to ν_2 can be orders of magnitude above normal Raman scattering efficiency; the emission at ν_2 also will be in a highly collimated beam. In theory, no dispersing device is needed for CARS. In addition, fluorescence and thermal background from heated samples, although of substantial difficulty in normal Raman scattering, is not a problem in CARS. CARS has been primarily used as a nonintrusive probe into hostile environments such as combustion flames or chambers (20,21). Hence, it should also be possible to use CARS as a monitor of gas phase chemistry, associated, for example, with methane activation.

Surface Enhanced Raman Spectroscopy (SERS)

It is generally very difficult to obtain the Raman spectrum of molecules adsorbed on planar or roughened, low-surface-area metals. However, Fleischman, Hendra, and McQuillan in 1974 measured with relative ease the Raman spectrum of pyridine on a silver electrode (22,23), which by the research of Jeanmaire and Van Duyne (24) and Albrecht and Creighton (25) in 1977 was explained to be the result of 10^4 to 10^6 enhancement of the Raman cross section of the adsorbed molecules. Since these pioneering studies, there has been an outpouring of theoretical and experimental work on SERS. Only with the most recent work has a clearer but still somewhat controversial concensus arisen concerning the theoretical explanation of SERS. Numerous in-depth reviews and otherwise independent articles have been published on the mechanisms that cause 10^4 to 10^8 enhancements in Raman intensities of adsorbed species. These discussions are summarized in this section with no attempt to present the reader with the basic electrodynamics or quantum chemical equations that are at the heart of this exciting development in Raman spectroscopy.

Generally, the observed effects in spectra obtained from SERS, compared to those from normal Raman spectroscopy, are as follows:

1. Vibrational bands of adsorbed molecules that were relatively weak or symmetry forbidden in the bulk may be active in SERS.
2. Excitation profiles (i.e., Raman band intensities as a function of excitation frequency) are substantially different for SERS.
3. Excitation light polarized in the plane of incidence (*p* polarization) generally produces higher SERS intensities than light polarized perpendicular (*s* polarization) to the plane of incidence.
4. The background is usually high for SERS of adsorbed species.
5. The SERS spectra generally are depolarized; normal Raman scattering spectra from adsorbed species contain highly polarized bands.
6. The intensity of SERS spectra is not necessarily a linearly increasing function of the concentration of the adsorbate.
7. Positions of bands are shifted from their normal bulk values, and as a function of excitation frequency for SERS.

The causes of surface enhanced spectra (SES) can be divided into two categories: electromagnetic, described by application of Maxwell's equations; and chemical, described by quantum chemistry. That substantial enhancement can be induced on roughed surfaces by electromagnetic phenomena is fairly well accepted (26,27). The formation of newly excitable states by the chemisorption of a molecule, one in which resonant Raman spectra would be excited using an otherwise nonresonant frequency of excitation, is also possible. However, it has been shown that the intensity of bands from SERS is up to 10^3 times greater for nonresonant excitation than for resonant excitation (28). The dependence of band intensity, I, in SERS upon fundamental parameters of the system under study is given by:

$$I(\nu) \propto \nu^4 |G_3|^2 R^2 \tag{2.13}$$

where ν is the frequency of the Raman line, G_s is an emission tensor that is dependent on surface geometry, and R is a reflection tensor that is dependent on the dielectric properties of the medium under investigation.

The geometrical factor in Equation (2.13) has been closely studied in SERS. For a flat surface, the enhancement of the Raman signal is minor, having possibly a range of ~4 to 30. For a microscopically roughened surface the Raman intensity will vary as a function of the fourth power of the size of the microscopic bumps. Because these qualitative statements are the result of electromagnetic treatment, the distance over which such enhancement develops can be considered large relative to mean interatomic separations. Such a long-range influence has been dramatically demonstrated, where SES have been observed for molecules located near colloidal particles (29).

An article by Jha (30) has given a comprehensive treatment of the electromagnetic and chemical effects that lead to SES. The chemical effect due to a short-range enhancement of the Raman cross section is attributed to modifications in the electronic and Raman polarizabilities of the adsorbed molecule. The quantum overlap of wave functions of the adsorbate and adsorbent, charge transfer between these molecules, and chemical bonding itself were shown in certain cases to significantly change the polarizability, α, of the adsorbate. This modification may also affect $\partial\alpha/\partial Q_k$, that is, the derivative of α with respect to the normal coordinates of the molecule, as a result of the interactions with the electronic states of the substrate. The enhancement expected to be caused by chemical effects is on the order of 10^2 to 10^3.

The SERS effect has been observed for the metals Ag, Au, Cu, Pd, In, Hg, Tl, Al, Li, Na, K, Ga, Zn, and Cd. These metals enable SES of adsorbed molecules to be obtained, but only under special conditions. For example, Zeman and Schatz (31) have discussed enhancements expected over these metals as a function of the excitation frequency and size of imperfections on the metal surfaces. Shifts in band positions from SER spectra of adsorbates on alkali halides, in addition to bands that are broadened as compared to their spectra in neat liquids, have been attributed to chemical effects. Oxidation of or impurities on a surface prior to adsorption of the trial molecules may prevent the observation of SES. However, as pointed out by Murray (32), pristine transition metal substrates should be able to be activated for SES. For the model particle shown in Figure 2.5, where the inner particle is Au having an outer shell of Ni of varying thickness, the magnitude of the enhancement of spectra as a function of excitation frequency is shown in Figure 2.6. An enhancement of 4×10^3 would be sufficient for observation of very small amounts of most adsorbates; additionally, other metals of importance in catalysis may be amenable to observation by SERS using sample modification and other enhancement "tricks." These SES are further discussed in Chapter 5.

Selection rules for SES of adsorbed molecules have been described (33–35). These selection rules are considerably different from those described for normal Raman spectroscopy and indicate that three types of bands exist in SES. The surface normal is arbitrarily described as the z direction with x and y parallel to the surface. The three band types belong to the tensor components α_{zz} (type 1); α_{xx}, α_{yy}, and α_{xy} (type 2); and α_{xz} (type 3). The basis for such a division of

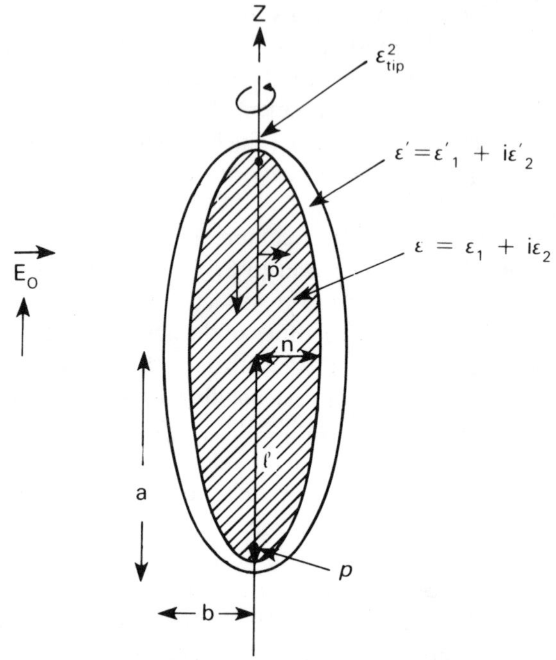

Figure 2.5. A model coated island. The inner prolate spheroid (shaded) has semimajor and semiminor axes l and *n*, respectively, and dielectric constant ε'. The semimajor and semiminor axes of the outer spheroid are *a* and *b*, and *p* is the tip radius of curvature. The background dielectric constant is ε_3. For the case of the prolate spheroids, at the resonance of interest the dipole moment *P* is directed along the long axis of the spheroid. The external field E_0 can be considered uniform if a $\langle\langle$ λ_0, the wavelength of the incident light. It is parallel to the long axis of the island here for simplicity (32). (Reprinted with permission from *The Journal of the Optical Society*, Copyright (1985), Journal of the Optical Society of America.

the tensor components and the fact that the Raman bands should scale in the sequence $\alpha_{zz} > \alpha_{xz}, \alpha_{yz} > \alpha_{xx}, \alpha_{yy}, \alpha_{xy}$ are natural consequences of the interaction of group theory, normal coordinate analysis, and classical electromagnetic theory.

Normal Raman Spectroscopy from Surfaces

The use of normal Raman (NR) spectroscopy to detect molecules adsorbed onto surfaces must rely on the maximization of experimental parameters, including scattering geometries, laser wavelength, monochromator and detector system,

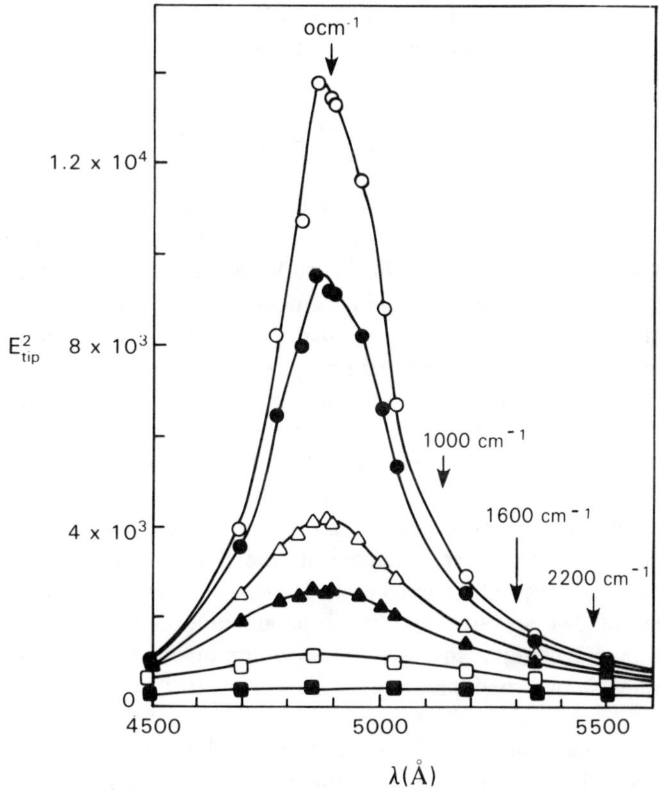

Figure 2.6. Calculated local electric field intensity at the tip of a set of two confocal prolate spheroids versus wavelength of incident light in vacuum. The inner ellipsoid is silver, and the outer spheroidal shell is nickel. The exciting laser line at 488 nm and various Raman-Stokes-shifted wavelengths are marked. The different curves (smooth lines drawn through calculated points) are values of the local field intensity for various average thickness *tl* to the outer gold layer. Open circles, *tl* = 0 nm; filled circles, *tl* = 0.045 nm; open triangles, *tl* = 0.19 nm; filled triangles, *tl* = 0.26 nm; open squares, *tl* = 0.45 nm; and filled squares, *tl* = 0.95 nm. The semimajor axis of the inner silver spheroid is taken to be 150 Å, and the ratio of its semiminor to semimajor axes is 0.31. The thicknesses *tl* above scale linearly with *l*, the semimajor axis of the inner silver spheroid (32).

and state of the substrate, to enable relatively intense Raman scattering intensities from monolayer species. Hallmark and Campion (34,36,37) have had significant success investigating how to proceed with such maximization, and, in the process, have displayed the utility of NR spectroscopy in the determination of molecular adsorption sites and symmetry on metal surfaces. Electromagnetic and/or chemical enhancements are not operative in NR spectra. For example, the induced dipole on a small spherical particle can be represented by:

$$\mu = [(\varepsilon_1(\nu) - \varepsilon_2)/(\varepsilon_1(\nu) + 2\varepsilon_2)]r^3E = \beta E$$

where $\varepsilon_1(\nu)$ is a complex, frequency-dependent dielectric function of the spherical particle, ε_2 is the ambient dielectric function, r is the radius of the particle, and E is the incident electric field of the laser. When Real $[\varepsilon_1(\nu_r)] \rightarrow 2\varepsilon_2 \neq 0$, a resonant condition develops, and enhancement in the scattering, where the intensity of the scattering is dependent on the magnitude of the induced dipole moment, becomes proportional to imaginary $[\varepsilon_1(\nu_r)]$. This situation accounts for limitations in the number of pure substrates on which SERS generally has been observed, that is, Au, Ag, and Cu, as these metals have high visible reflectivities, and their imaginary $[\varepsilon_1(\nu_r)]$ are very small. In the case of relatively low-reflectivity metals, such as Ni, Co, Fe, Pt, Pd, and so on, the imaginary $[\varepsilon_1(\nu_r)]$ values are quite large. In some cases, an increase in the surface area of these substrates with a concomitant increase in the number of scattering centers has been required to produce detectable Raman signals. However, continued development of NR spectroscopy is of paramount importance because it is the only known technique that can be used for quantitatively observing adsorbed species in a broad pressure range, from 10^{-10} to above atmosphere conditions.

As in the Raman analysis of bulk samples and supported phases, Raman activity for adsorbed molecules is determined by molecular polarizability components, which transform in the same way as the binary Cartesian components x^2, y^2, z^2, xy, xz, and yz. For an isolated adsorbed molecule, however, the presence of the adsorbent will change the symmetry of the adsorbate scattering center (38). This scattering center includes the adsorbent surface, which could be considered a plane containing adsorbate sites whose symmetry is dependent on the surface atomic geometry. Additionally, although image dipole theory has been extended to address selection rules for Raman activity (39,40), imperfect image screening produced by poor conductors precludes definitive selection rules. Rather, a set of "propensity rules" (34) is produced that will predict relative intensities of Raman bands.

A change in the symmetry of a molecule due to adsorption could produce Raman activity for otherwise inactive vibrational modes. Moskovits (41) has presented a detailed theoretical analysis of such an adsorbed molecule using image dipole formulism. For an isolated point dipole with the z direction normal to the surface, his results show that the intensity, I, of scattered s and p polarized light for a perfect conductor would be:

$$I_{ss} = 0$$

$$I_{sp} = I_{ps} = 0$$

$$I_{pp} \propto \{\nu^4[\alpha_{zz}(1 + r_{pi})\,(1 + r_{pi})]^2 \sin^2 \phi \sin^2 \phi\}$$

where r_{pi} and r_{pr} are the Fresnel coefficients of incident and scattered polarizations, and ϕ and θ are incident and scattering angles of the incident excitation with respect to the surface and adsorbed molecule. In the case of an imperfect conductor, and considering the Cartesian axis with z normal to the surface, yz in the plane of incidence (p), and x perpendicular to the plane of incidence (s), the trend in activity is the same as for SERS—that is, the zz component would be the most intense, xz and yz are less intense, whereas xx, yy, and xy are extremely weak to forbidden. Intuitively, such polarization selectivity implies that a z-oriented field can induce dipoles in the x, y, and z directions. However, because the metal surface is located within the x and y plane, the x and y components of the dipole will be screened. The x and y–oriented field can also induce dipoles in x, y, and z directions, with only the xz and yz components containing possibly insignificant screening or appreciable scattered intensities.

Symmetry reduction upon adsorption will produce a molecular symmetry that is a subgroup of the molecular point group. For benzene, point group D_{6h}, the number of correlation possibilities to subgroups is 29 in number. However, as described by Hallmark and Campion (37), adsorption with the benzene C_6 axis normal to a planar surface would produce an effective symmetry of C_{6v}. In the case of the C_2 axis of benzene oriented parallel to a "trough" of a (110) surface, the effective symmetry would be C_{2v}. For an array of adsorbed, two-dimensional surface species, the effective symmetry of the array would be the factor group symmetry of the original species.

The rapid damping of the incident electromagnetic wave over a short distance within a metal suggests that the electric field gradient at its surface would be significantly large. In this case, contributions from higher order terms in the polarizability expression (Equation 2.4) could be important. In particular, the quadrupole polarizability tensor (β) has been suggested to be important in Raman scattering from surfaces (42). The comparison of spectra of an adsorbate on different faces of a metal could possibly distinguish the influence of quadrupole terms on the generation of new Raman bands.

Resonant Raman Scattering

The application of normal Raman (NR) spectroscopic techniques provides molecule-specific information about the ground vibrational and electronic states of gases, liquids, and solids. In contrast, when the frequency of the exciting laser is close to and/or within an electronic absorption band of a molecule under investigation, the excited electronic and corresponding vibrational states of that molecule can be monitored. This type of Raman scattering has been called resonant Raman scattering (RRS). It has been shown that specific parts of complicated molecules can be probed by use of suitable excitation frequencies. Such a possibility also occurs for molecules in solution, where detuning of the laser frequency away from absorption bands of the solvent can allow the detection of solvated molecules, or part of a solvated molecule, independent of the solvent. The early development of excited state vibrational spectra normally included only continuous wave excitation and spectral observation. However, more recent development of pulsed laser sources and synchronously time-gated detection systems has enabled the observation of transient species that are created during pulsed excitation. Such observation implies that measurements of chemical kinetics are possible, in addition to normally acquired information such as molecular symmetry, bonding distances, perturbations due to interactions with solvent or other nearby molecules, and molecular concentrations. Pulsed or transient Raman experiments have gained the descriptive title of time resolved vibrational spectroscopy (TRVS) or time resolved resonance Raman (TR^3) spectroscopy.

The amount of information published on resonant Raman spectroscopy has not approached that published in normal spectroscopy. As a result, libraries on resonant spectra to which spectra of samples can be compared are incomplete. Such a deficiency, albeit temporary in nature, increases the difficulty of spectral interpretation. In addition, specialized equipment and theory need to be considered for TR^3 spectroscopy, making it considerably more difficult to apply than normal Raman spectroscopy. Control of the excitation source, both power and temporal, is important in data acquisition because nonlinear effects should be avoided while excitation spectral features are observed.

To begin the theoretical discussion of resonant Raman effects, it may be helpful to consider an energy diagram that displays the fundamental difference between NR and RR excitation (see Figure 2.7). In the case of NR spectroscopy, the excitation is far from resonance, so transitions of electrons creating the Raman scattering can be considered to involve a virtual state energetically removed from the first excited electronic levels. On the other hand, the resonant case involves the excitation of electrons from their ground level to an excited level, having either a continuum or discrete vibrational levels.

Spectral characteristics of NR and RR spectra can be compared as follow: In the NR case: (1) Rayleigh scattering is intense while the fundamentals are relatively weak; (2) overtones and combination bands contain weak intensities in comparison to the fundamentals; (3) scattering intensities vary as a function of the incident frequency to the fourth power; (4) the relative intensity of stokes to anti-stokes scattering can be calculated by use of the Boltzmann factor; (5)

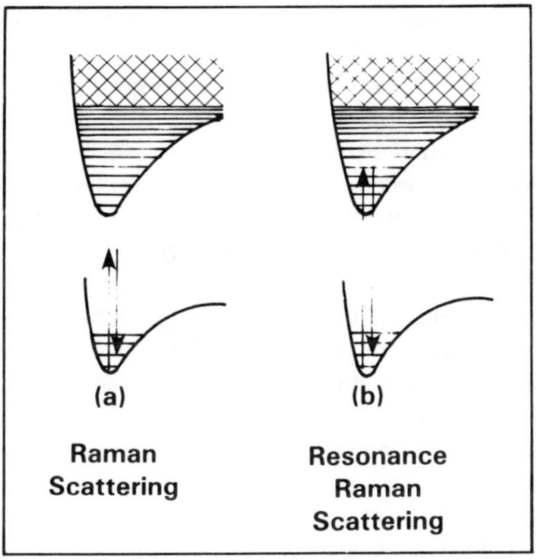

Figure 2.7. Classifying Raman scattering according to laser frequency and molecular absorption band. (a) The incident laser frequency is far from resonance with any real electronic transitions, so normal Raman scattering is observed. (b) The incident laser frequency is in the region of discrete levels of a single electronic intermediate state; so resonance Raman scattering is observed.

scattering is instantaneous; (6) quenching of the scattering intensities from a particular molecule cannot be accomplished without destruction of the molecular symmetry or identity; and (7) scattered band envelopes are relatively broad and independent of the incident frequency. In the RR case: (1) intensities of the fundamentals may be as large as the Rayleigh line; (2) overtone and combination bands can have intensities larger than and/or equal to the fundamentals; (3) relative intensities of the overtones may be erratic; (4) the number of overtone and/or combination bands is limited by symmetry selection rules; (5) widths of the overtones are dependent on excitation frequency within any particular molecule, and also dependent on fundamental physical characteristics of the electronic levels of the molecule; (6) scattering intensities are not dependent on the fourth power law, but are somewhat unpredictable; (7) the scattered band envelope is dependent on the excitation frequency; and (8) the scattering time is slow, as compared to NR scattering, usually in the range of 10^{-12} to 10^{-5} second.

A number of reviews have theoretically treated resonant Raman scattering (43–46). It is not the author's intention to address this theory thoroughly here, but readers can improve their understanding of resonant phenomena by reviewing this literature. The generalized theory described in previous sections of this chapter accounts for resonant scattering; so only some parts of resonant theory will be addressed.

In terms of the temporal characteristics of resonant scattering, the Raman emission is delayed relative to NR emission because of the discrete lifetimes of excited states. During population of an excited state with electrons, and especially in the case of condensed systems, a considerable number of channels exist in which the energy of the excited electron can be lost. These channels relieve the electrons in the excited state of some of their energy and/or change their lifetime within an excited state.

The direct product terms in Equation (2.11) are called vibrational overlap terms, the squares of which are called Frank-Condon factors. A more comprehensive expression than Equation (2.11) can be found in an article by Rousseau, Friedman, and Williams (47); their notation is more explicit in showing that the excited state quantum mechanical notation also contains a term related to the width of the excited state. In addition, the mixing of electronic states, as expressed by Rousseau et al. (47), causes distortion of the excited states and the initial states such that extra intensity is "mixed" into vibrational modes that are otherwise weak. The structure of a resonant spectrum is dependent on whether the electron promoted into an excited state is in a discrete vibrational/rotational level. Because these levels are separated by a relatively small amount of energy as compared to the absolute energy in the excitation laser, a small change in excitation energy can change dramatically the structure of the resonant spectrum as a result of excitation into a different vibrational/rotational level.

Quantitation of RR spectral results can be accomplished in much the same manner as quantitation of NR spectral results. Usually, an internal standard can be added to the molecular system under investigation. Other methods of quantitation are presented in Chapter 6. For a system with only a single electronic state that predominates in the resonance effect, the intensity of a Raman band in a preresonance region can be written as:

$$F_a^2(\nu_s) = \nu_s^4(\nu_e^2 + \nu_{ex}^2)^2/(\nu_e^2 + \nu_{ex}^2)^4$$

where ν_e is the frequency (in wavenumbers) of the transition from ground to excited state. For the case of more than one electronic level involved in the resonance effect, the intensity can be expressed as:

$$F_b^2(\nu_s) = 4\nu_s^4(\nu_{e1}\nu_{e2} + \nu_{ex}^2)^2/[(\nu_{e1}^2 - \nu_{ex}^2)^2(\nu_{ex}^2 - \nu_{ex}^2)^2]$$

where ν_{e1} and ν_{e2} are the frequencies of the transitions from the ground state to the two electronic levels, 1 and 2.

REFERENCES

1. Raman, C. V. and K. S. Krishnan, *Nature 121,* 501 (1928).
2. Wilson, E. B., Jr., J. C. Decius, and P. C. Cross, *Molecular Vibrations, The Theory of Infrared and Raman Vibrational Spectra* (McGraw-Hill, New York, 1955).
3. Long, D. A., *Raman Spectroscopy* (McGraw-Hill, New York, 1977).
4. Aikins, P. W., *Molecular Quantum Mechanics* (Clarendon Press, Oxford, 1970).
5. Colthup, N. B., L. H. Daly, and S. E. Wiberley, *Introduction to Infrared and Raman Spectroscopy* (Academic Press, New York, 1964).
6. Wheatley, P. J., *The Determination of Molecular Structure* (Claredon Press, London, 1968).
7. Nakamoto, K., *Infrared and Raman Spectra of Inorganic and Coordination Compounds* (Wiley, New York, 1978).
8. Barron, L. D., and A. D. Buckingham, *Ann. Rev. Phys. Chem. 26,* 381 (1975).
9. Long, D. A., and L. Stanton, *Proc. Roy. Soc. A318,* 441 (1970).
10. Christie, J. H., and D. J. Lockwood, *J. Chem. Phys. 54,* 1141 (1971).
11. Jones, W. J., and B. P. Stoicheff, *Phys. Rev. Lett. 13,* 657 (1964).
12. Maker, P. D., and R. W. Terhune, *Phys. Rev. 137A,* 801 (1965).
13. Cotton, F. A., *Chemical Applications of Group Theory* (Interscience, New York, 1963).
14. Jaffe, H. H., and M. Orchin, *Symmetry in Chemistry* (Wiley, New York, 1965).
15. Herzberg, G., *Molecular Spectra and Molecular Structure,* vol. 2., *Intrared and Raman Spectra of Polyatomic Molecules* (Van Nostrand, New York, 1945).
16. Carter, R. L., in *Infrared and Raman Spectroscopy, Part A,* ed. E. G. Brame, Jr. and J. Grasselli (Marcel Dekker, New York, 1976), p. 71.
17. Young, E. S., *J. Mol. Spectrosc. 53,* 379 (1974).
18. Stone, J., *J. Chem. Phys. 69,* 4349 (1978).
19. Koszykowski, M. L., L. A. Rahn, R. E. Palmer, and M. E. Coltrin, *J. Phys. Chem. 91,* 41 (1987).
20. Hall, R. J., and A. C. Eckbreth, in *Laser Applications,* ed. J. F. Ready and R. K. Erf, vol. 5 (Academic Press, New York, 1984), pp. 213–319.
21. Pealat, M., P. Bouchardy, M. Lefebure, and J.-P. Taran, *Appl. Opt. 24,* 1012 (1985).
22. Fleischman, M., P. J. Hendra, and A. J. McQuillan, *Chem. Phys. Lett. 26,* 163 (1974).
23. McQuillan, A. J., P. J. Hendra, and M. Fleischman, *J. Electroanal. Chem. 65,* 933 (1975).
24. Jeanmaire, D., and R. P. Van Duyne, *J. Electroanal. Chem. 84,* 1 (1977).
25. Albrecht, M. G., and J. A. Creighton, *J. Am. Chem. Soc. 99,* 5215 (1977).
26. Wokaun, A., *Mol. Phys. 56,* 1 (1985).
27. Metiu, H., and P. Das, *Ann. Rev. Phys. Chem. 35,* 507 (1984).
28. Stacy, A. M., and R. P. van Duyne, *Chem. Phys. Lett. 102,* 365 (1983).
29. Abe, H., W. Schulze, and B. Tesche, *Chem. Phys.47,* 2200 (1980).
30. Jha, S. S., *Surf. Science 158,* 190 (1985).
31. Zeman, E. L., and G. C. Schatz, *J. Phys. Chem. 91,* 634 (1987).
32. Murray, C. A., *J. Opt. Soc. Am. 2B,* 1330 (1985).
33. Moskovits, M., and J. S. Suh, *J. Phys. Chem. 88,* 1984 (1984).
34. Hallmark, V. M., and A. Campion, *J. Chem. Phys. 84,* 2942 (1986).
35. Creighton, J. A., *Surf. Science 158,* 211 (1985).
36. Hallmark, V. M., and A. Campion, *Chem. Phys. Lett. 110,* 561 (1984).
37. Hallmark, V. M., and A. Campion, *J. Chem. Phys. 84,* 2933 (1986).
38. Nichols, H., and R. M. Hexter, *J. Chem. Phys. 75,* 3126 (1981).
39. Hexter, R. M., and M. G. Albrecht, *Spectrochim. Acta 35A,* 233 (1979).
40. Nichols, H., and R. M. Hexter, *J. Chem. Phys. 73,* 965 (1980).

41. Moskovits, M., *J. Chem. Phys. 77,* 4408 (1982).
42. Sass, J. K., H. Neff, M. Moskovits, and S. Holloway, *J. Phys. Chem. 85,* 621 (1981).
43. Tang, J., and A. C. Albrecht, in *Raman Spectroscopy Theory and Practice,* ed. H. A. Szymanski, vol. 2 (Plenum Press, New York, 1970).
44. Jacon, M., in *Advances in Raman Spectroscopy,* ed. J.-P. Mathieu, vol. 1 (Heydon and Son, London, 1973).
45. Jan Labeke, D., M. Jacon, M. Berjot, and L. Bernard, *J. Raman Spectrosc. 2,* 219 (1974).
46. Behringer, J., in *Molecular Spectroscopy,* ed. R. E. Barrow, D. A. Long, and D. J. Millen, vol. 3 (The Chemical Society, Burlington House, London, 1975).
47. Rousseau, D. L., J. M. Friedman, and P. F. Williams, in *Raman Spectroscopy of Gases and Liquids,* ed. A. Weber, (Springer-Verlag, New York, 1979).

3 Instrumentation

INTRODUCTION

The ease of varying the type of instrumentation used to acquire Raman spectroscopic information from catalysts is a significant benefit in addressing specific questions about adsorbed species and catalyst chemical/physical structure. This convenience is due partly to the modular nature of Raman instrumentation, with only the excitation source and the detector constant elements in any Raman system. Among the parameters that can be varied to match the experimental design are: the scattering geometry; the sampling configuration; the controlled environmental sampling cells; the monochromator or interferometer; the excitation frequency, and whether the laser is continuous or pulsed; the detector type and amplification systems; data retrieval and handling; and the polarization dependency. Catalysts can be investigated under conditions that closely resemble their application conditions without any need for ultrahigh vacuum conditions. Interferences in spectral acquisition, if a potential problem, can be eliminated or decreased to an insignificant level prior to data acquisition, by methods that are discussed below.

This chapter will describe instrumentation possibilities first by defining the essential parts of a Raman data system and then by discussing refinements that can be used to produce good-quality molecular vibrational information on catalysts. The many options available for the application of laser Raman spectroscopy to catalysis will be presented. These options are increasing at an astonishing rate, and in many cases owe their discovery and initial application to fundamental research in areas other than catalytic research.

RAMAN SPECTROMETERS

A block diagram of a typical laser Raman spectrometer is shown in Figure 3.1. Particular elements of this system include a laser, collimating lenses or mirrors, sample cell and positioner, input optics, monochromator, detector, amplification electronics, computer interface and computer, and spectral display system. Operation of such a spectrometer entails the excitation of the vibrational modes of a sample by laser irradiation and the collection of Raman scattered radiation. The monochromator contains dispersive elements, generally holographic grat-

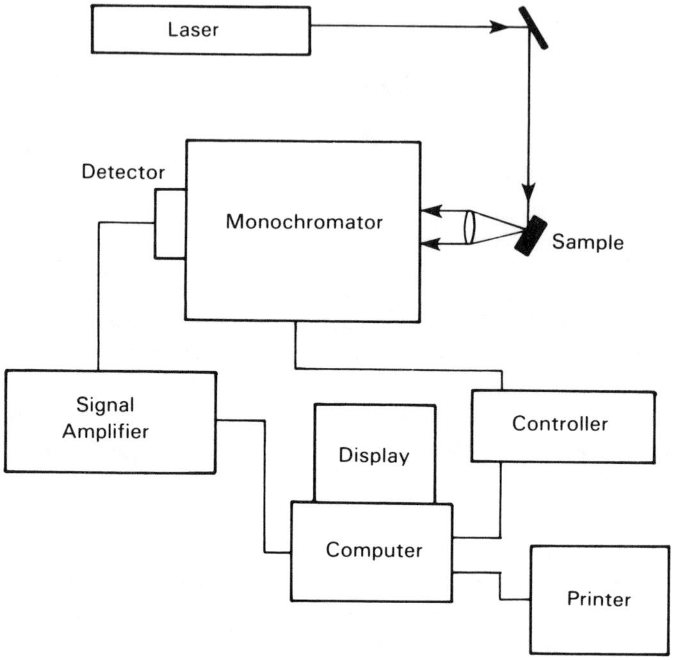

Figure 3.1. Typical components in a laser Raman spectrometer used for normal Raman scattering.

ings, whose number depends on the amount of discrimination needed for separating the highly intense Rayleigh scattering from Raman scattering. Figure 3.2 shows a dispersive Raman spectrometer containing a three-grating turret design with a microprocessor-controlled spectral repositioning or scanning system for multichannel or single-channel detection.

The intensity of Raman scattering is on the order of 10^7 to 10^8 times less than Rayleigh scattering. The maximum frequency resolution of commercially available dispersive spectrometers is about 0.5 cm^{-1} ([10,000/(frequency in cm^{-1})] = wavelength in micrometers; and 1 eV = 8088 cm^{-1}). The components of the spectrometer shown in Figure 3.1 are typical of those used in normal Raman (NR) spectral acquisition and in surface enhanced Raman spectroscopy (SERS). More recent methods used to spectrally separate the frequency response of the sample under investigation include Fourier-transform Raman spectroscopy (FTRS) and Hadamor spectroscopy (HS). The experimental equipment associated with FTRS will be discussed later in this chapter.

The capabilities of Raman spectroscopy for catalysis research are special. To indicate this uniqueness, Table 3.1 compares Raman spectroscopy with two techniques commonly used in catalysis research. As mentioned in Chapter 1,

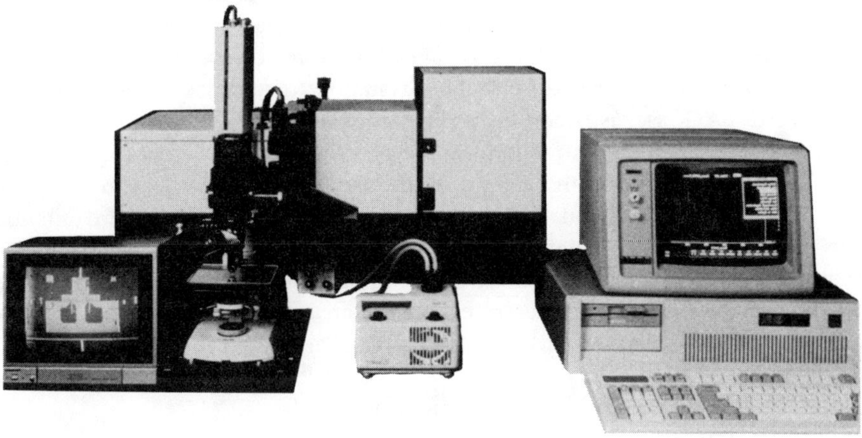

Figure 3.2. Instrumentation for performing normal Raman spectroscopy (compliments, Spex Industries, Inc.).

Raman spectroscopy is considered to be as flexible as (and possibly more flexible than) the traditionally used infrared spectroscopy techniques. The flexibility of Raman spectroscopy and the ability to monitor it under realistic reaction conditions are noteworthy advantages.

The type of laser used generally depends on the experimental requirements

Table 3.1. Comparison of vibrational spectroscopic techniques.

	INFRARED	ELECTRON ENERGY LOSS	RAMAN	SURFACE ENHANCED RAMAN
Spectral range	30–4000 cm^{-1}	100–4000 cm^{-1}	10–4000 cm^{-1}	50–4000 cm^{-1}
Spectral resolution	0.1–10 cm^{-1}	30–100 cm^{-1}	1–10 cm^{-1}	1–10 cm^{-1}
Sensitivity	0.01 monolayer, < pmole	0.01 monolayer	<0.1 monolayer, < pmole	0.01 monolayer
Selection rules	Known	Known	Known, estimated	Partially known
Sampling	Any type sample, spectral interference at <1000 cm^{-1}	Smooth surface	Any type sample	Roughened surface
Conditions	Varied, vacuum sometimes	Vacuum	No normal restrictions	No normal restrictions

of the research. There are a number of manufacturers of lasers; the types most often employed in catalysis research include continuous or mode-locked argon ion, argon-krypton, and helium-neon. These lasers can be coupled to a dye laser, thereby increasing the range of visible excitation frequencies. In addition, the pulsed Nd-YAG laser is becoming a standard workhorse in Raman studies. The available laser frequencies range from the near-infrared to the near-UV region, in many cases allowing the tuning of an excitation source to a region that maximizes the Raman signal for a minimal amount of material and excitation power. Laser emission generally is linearly polarized; such polarization implies that the electric polarization vector of the light incident upon a sample can be easily aligned in the particular direction required for the experiment. Linear polarization also causes orientation-dependent changes in the intensity reflected by gratings in the monochromator; so the collection input optics for a monochromator should include a polarization scrambler.

Sample Cells and Sampling Techniques

Sample cells and/or the positioner shown in Figure 3.1 for application to gas, liquid, or solid samples are not specialty items, but are available from vendors of laser Raman instruments. In general, the designs of sample holders used to facilitate the acquisition of Raman spectra of solids without control of environments above the sample are relatively simple, depending only on geometrical considerations of the spectrometer; so such sample holders will not be discussed here. Specialty cells used for the acquisition of Raman data from samples under controlled conditions are not generally available from vendors, but normally are designed and/or constructed by experimentalists, taking into consideration specifications of the experiments and the samples. As early as 1971, Kiefer and Bernstein (1) described a cell specially built to control the environment of a solid sample during Raman data acquisition. Brown, Makovsky, and Rhee (2) used a modification of this design, in which the sample was rotated by a motor to obtain some of the first Raman data on catalysts. The rotation of samples now is practiced routinely in the acquisition of such data. Such rotation is necessitated by the otherwise-possible photodegradation of the sample with the intensely focused laser beam. Rotation of samples is required especially with colored samples and with excitation in the visible frequency range, to ensure the acquisition of sample-representative spectra. An alternative to sample rotation is optically controlled movement of the laser beam on the sample surface (see below); the sample does not have to rotated in this case.

Cheng, Ludowise, and Schrader (3) described a controlled-atmosphere cell that included a rotating sample design for in situ studies of catalysts. Makovsky, Diehl, and Stencel (4) used a design similar to that except that the components

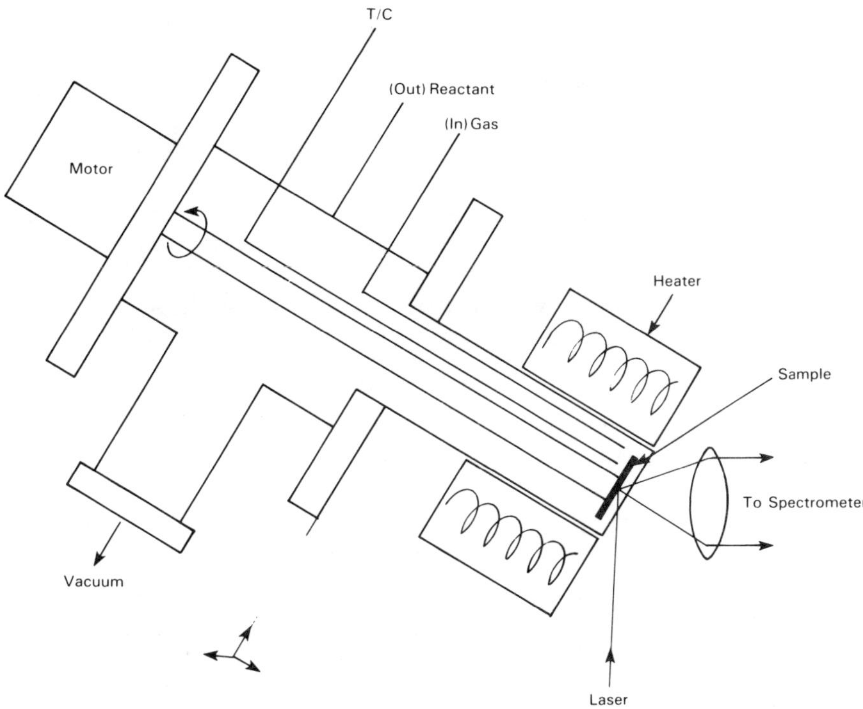

Figure 3.3. A ultrahigh-vacuum-compatible cell that is used for treating catalysts at high temperatures. The catalyst wafer is rotated during laser illumination.

of the cell were UHV-compatible (see Figure 3.3). This design contains a motor-driven rod extending down the center of a graded glass–metal seal and quartz sample cell. The end of this cell contains a quartz optical flat through which the laser light impinges upon the sample. Raman radiation is collected by the input optics of the monochromator. The quartz cell is surrounded at its end by a sample heater capable of temperatures as high as 700°C; the temperature of the catalyst is monitored by a thermocouple mounted via a pressure-sealed metal flange and extending to a position adjacent to and behind the sample. Generally, catalysts studied with this cell were pressed into thin 13-mm-diameter wafers and attached to the end of the rotating rod by a centering/screw mechanism. This mechanism secured the orientation of the catalyst perpendicular to the axis of rotation. Flanges at the motor end of the sample are UHV-compatible, containing Cu-knife-edge seals; the flange through which rotation was accomplished included a ferro-fluid-type seal. The cell is evacuable by a turbomolecular pump to an ultimate vacuum of approximately 10^{-9} bar. During exposure of catalysts to various gaseous mixtures with the catalyst at an elevated temperature, gate valves

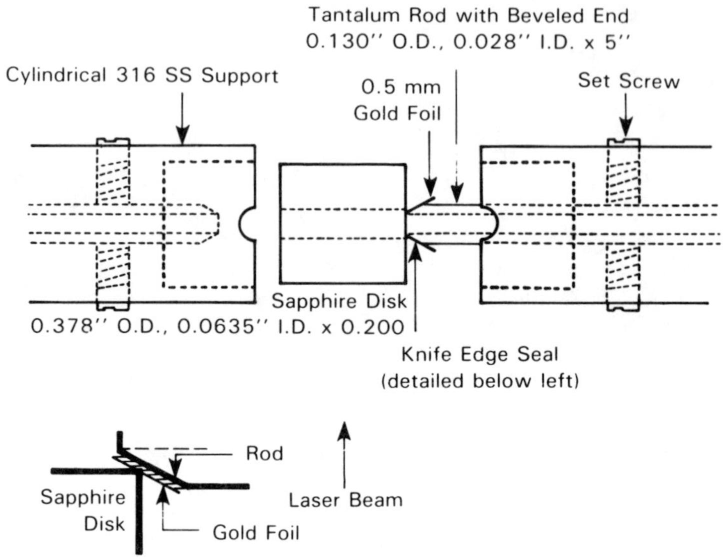

(a)

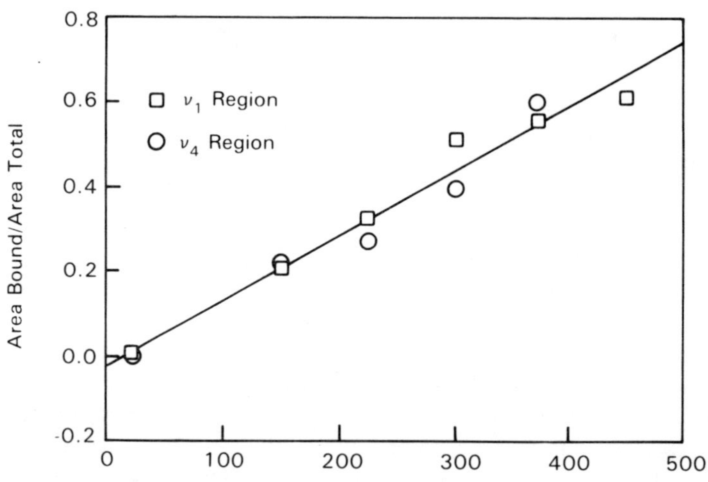

(b)

Figure 3.4. (a) An expanded view of the optical "window" region of a high temperature cell, illustrating how seal contact is made. For purposes of viewing, the right-hand Ta rod is shown protruding. When the cell is assembled, this Ta rod is recessed (as the left-hand Ta rod is in the figure). The lower insert expands this contact seal region still more (7). (b) The amount of contact-ion pairing of NO_3^- in aqueous $Zn(NO_3)_2)$ increasing linearly with temperature (7).

between the cell and the turbomolecular pumps were closed. Spectra have been obtained with reactant gas in the cell, with the sample under vacuum, and with the sample at elevated or ambient temperatures.

Other sample cells that are used in the acquisition of Raman spectra of catalysts and catalyst surfaces and can be used in conjunction with other surface-sensitive techniques have been described (5,6). For liquid samples, Spohn and Brill (7) recently described a cell that was usable at temperatures as high as 500°C and at pressures as high as 35 MPa. The design of this cell is shown in Figure 3.4, along with data on the amount of contact-ion pairing of NO_3^- in aqueous solutions of $Zn(NO_3)_2$. Such contact-ion pairing is a direct measure of the equilibrium distribution of NO_3^- ions as a function of temperature and displays changes in interactions between constituents in solution phases. The optical, sampling region of this cell consists of a polished single-crystal sapphire capillary disk with a polished inner bore that is filled with the liquid sample. The sapphire is supported between tantalum rods, and the sapphire disk is sealed against the tantalum rods by means of a thin gold washer. The sample position is heated by resistive heating cables, and the outer section of the cell is simultaneously cooled to allow its positioning in the spectrometer system while the sample is at high temperature. Also, Lin, Mann, and Vickers (8) described a liquid cell that used standard chromatographic fittings as connections to a windowless flow cell.

Use of the above-described sampling cells usually has entailed Raman scattering angles of around 60° to 90°. These angles allow great flexibility in the design of cells and enhance their coupling to a monochromator. Other sampling geometries, such as those in Figure 3.5, have been used in the acquisition of Raman data from catalysts. The 180° backscattering geometry in Figure 3.5a has been used routinely in an instrument called a Raman microprobe. The laser beam in this instrument is finely focused onto the sample to spots as small as one micrometer. Computerized movement of a sample stage and observation of regions on the sample emitting Raman radiation at specified frequencies are used to map molecular entities on the sample surface. Some Raman spectra of catalysts discussed in Chapter 4 were obtained with such microprobe instrumentation. Other areas of application of Raman microprobe spectroscopy include ferrography, mineral analysis, corrosion, and polymer microspectroscopy. The use of 180° backscattering does not allow the polarization vector of the incident light to be rotated normal to the sample surface. Although such a decrease in a degree of freedom has not been found to be hindrance in the study of finely divided powders, the study of molecules adsorbed on smooth metal and powdered samples could be adversely affected.

An alternate approach to rotating a sample during Raman data acquisition is to rotate the optics handling the incident laser light, thereby causing the light to move in a cyclical pattern on a stationary sample. Zimmerer and Kiefer (9) and Kiefer (10) described the optics necessary for this type of rotating lens experi-

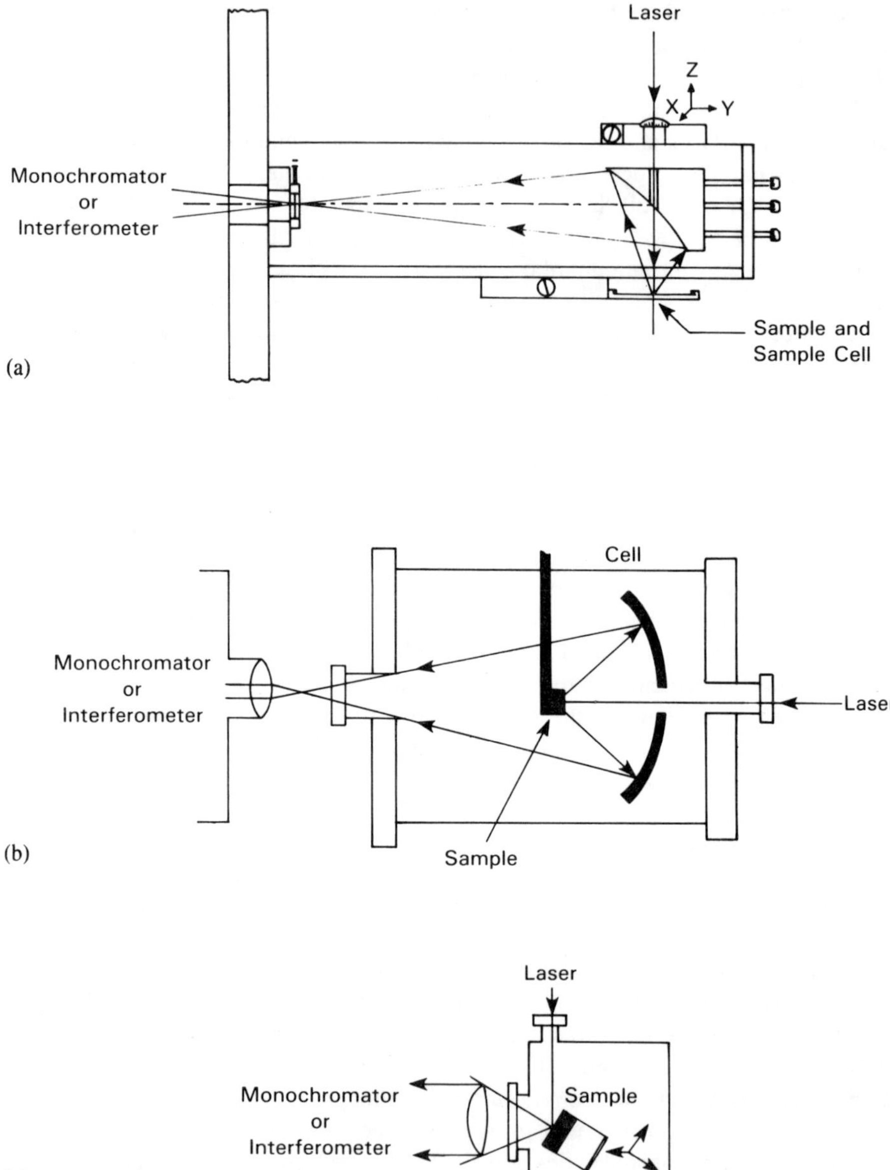

Figure 3.5. Some sample illumination geometries for Raman spectroscopy. Configurations (a) and (b) involve 180° back scattering, whereas (c) involves 90° scattering.

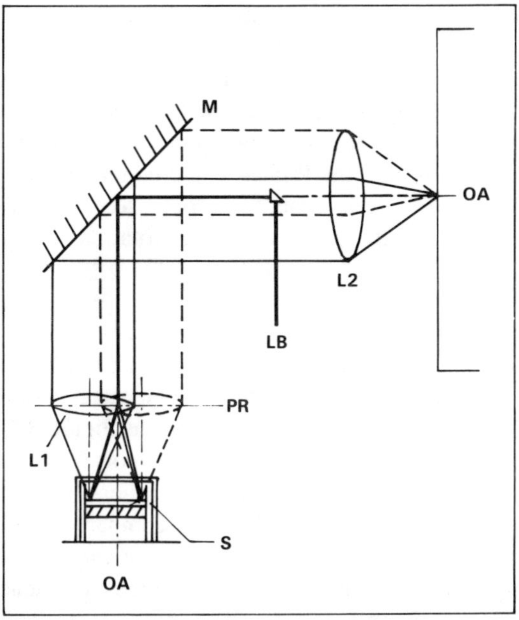

Figure 3.6. Rotating lens system, in which the laser focal spot on the sample is moved instead of rotating the sample. *LB:* laser beam; *OA:* optical axis; *M:* planar mirror, *L*1: condenser lens; *L*2: telescope lens; *S:* sample surface; *PR:* plane of rotation (9,10).

mentation; their system is shown in Figure 3.6. A column of laser light (*LB*) is directed toward a sample (*S*) along the optical axis (*OA*) of the spectrometer (*SP*) by use of a prism. This light is deflected by a planar mirror (*M*) to the direction of the sample and condenser lens (*L*1). The condenser lens is mounted off-axis with respect to the incident laser beam and is attached to a device that rotates it at speeds of up to 1000 rpm. Light that is focused onto the sample by the rotating lens is incident on the sample, displaced from the lens by a distance equal to the focal length of the lens, at the center of the lens. Because the lens is rotated, its focal point rotates in a circular manner, scribing a circle (*S*) of light on the sample.

Scattered Raman radiation is collected by the rotating condenser lens and directed toward the spectrometer by the mirror. At the entrance to the spectrometer there is a telescope lens (*L*2), which focuses the light to a point that remains stationary on the spectrometer slit. Hence, the intensity at the slit is constant and not perturbed by the rotation of the condenser lens.

The advantages of using the rotating lens systems include an ease of sample preparation not possible with rotating samples, as well as decreased difficulty in the design characteristics of the sampling cell in which the sample is to be placed. With sample rotation, it is relatively difficult for some samples to be pressed into self-supporting wafers to be attached to a rotational feed-through. However, powders are easily sampled with a rotating lens design. Precise alignment of the rotational feed-through and a self-centering sample holder are required for the use of a rotating sample configuration. Otherwise, and in the case of imperfect sample wafering, the Raman scattered radiation incident on the slit of the spectrometer will precess around the optic axis, causing modulation of the Raman intensity at the frequency of sample rotation.

Detectors

Detectors used in Raman spectroscopy include photomultipliers and multichannel analyzers using Si and/or Ge photodiodes. The spectral responses of a photomultiplier and Ge and Si photodiodes are shown in Figure 3.7. The sensitivity of a detector is measured in terms of its background noise, its quantum efficiency, and the frequency range of its use. Photomultipliers are known to have an extraordinarily low background noise level as compared to Ge and Si photodiodes. However, as indicated in Figure 3.7, the quantum efficiency of photomultipliers is only about one-fourth of the photodiodes' efficiency; and the quantum efficiency of multichannel detectors is approximately one-half of that for photomultipliers. This efficiency is best for diodes in the red end of the visible region and toward the near infrared, whereas the photomultiplier has generally constant efficiency in the mid-visible to near-UV region.

Photomultiplier tubes were the standard detectors used in laser Raman spectrometers; but the use of multichannel analyzers has increased because of the increased speed—a consequence of their multiplexing nature—with which spectra can be acquired, as compared to the speed with the use of photomultipliers.

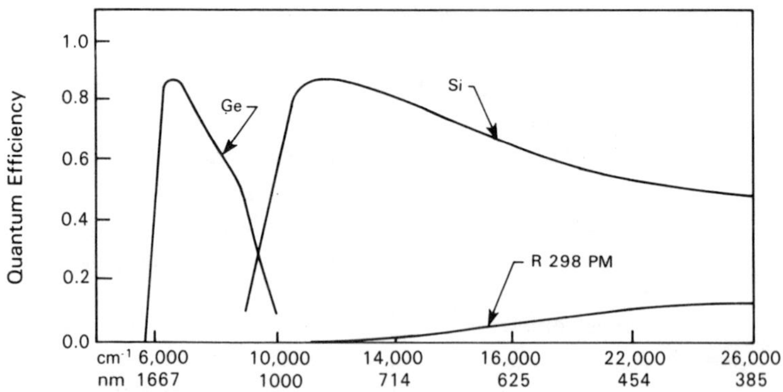

Figure 3.7. Responses of typical detectors used in Raman spectroscopy: Germanium photodiode (Ge), silicon photodiode (Si), and a photomultiplier.

However, the maximum spectral resolution that is possible with multichannel analyzers is inferior to that achieved with photomultipliers.

Pulsed Raman Data Acquisition

The acquisition of Raman data from catalysts using nanosecond (10^{-9} second) and picosecond (10^{-12} second) pulsed lasers is now possible (11). Such pulsed experiments enable the study of transient species that may be important in catalytic reactions. Components of a nanosecond pulsed Raman system are shown in Figure 3.8. This system is similar to that used by Asher, Johnson, and Murtaugh (12) to monitor the Raman spectra of constituents in liquid mixtures/solutions at very low concentrations. The lasers that can be used in this arrangement include Nd-YAG, Ar^+, and a flash-lamp excited dye laser. The laser wavelength impingent upon the sample can be varied over the approximate range 200 to 750 nm; with the use of an H_2 shift cell, the range of excitation could be extended to cover 185 to 880 nm (13,14). The fundamental lasing wavelength of the YAG laser is at 1064 nm, and this emission can be frequency doubled or tripled to produce lasing at 532, 355, and 266 nm. Harmonics of the fundamental lasing frequencies also can be used, for example, at 213 nm (fifth harmonic). Extension of the excitation frequency to the UV region enhances the ability to tune into the electronic resonance absorption effects associated with the appro-

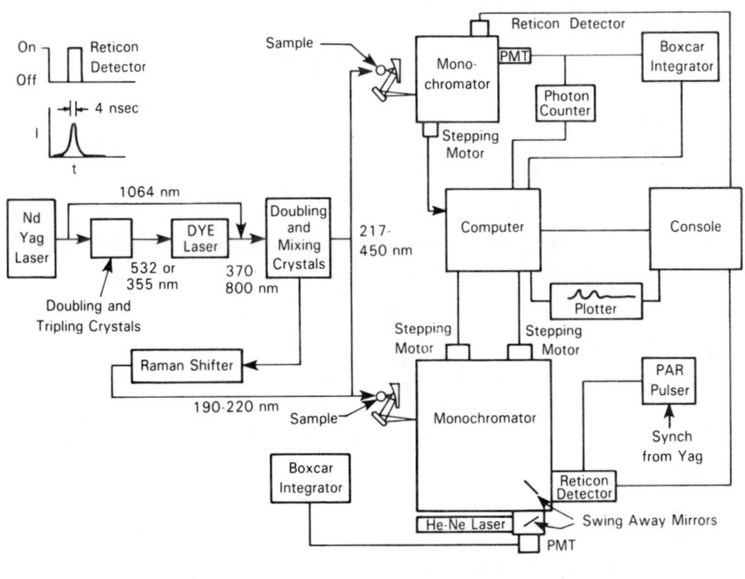

Figure 3.8. Block diagram of a wide frequency, pulsed Raman spectrometer for resonance Raman scattering (12).

priate excitation of the sample. In the case of the spectrometer depicted in Figure 3.8, the pulse length of the laser can be as small as a couple of nanoseconds.

The components of a picosecond pulsed Raman system are shown in Figure 3.9 (15). Many of these components are similar to those shown in Figure 3.8. The mode-locked Nd-YAG oscillator consists of a passive mode-locking system in which a saturable dye absorber is placed within the resonator cavity. Periodic losses in output power density (fluence) are introduced by an intracavity acousto-optic modulator. Frequency beating of these locked modes produces a series of high-intensity picosecond pulses that exit the resonator cavity. Outside the cavity, an avalanche-transistor extracts one picosecond pulse from the series of pulses, and this pulse is amplified by a double-pass amplifier. The frequency of these selected pulses can vary according to experimental design, but laser fluence recovery and power stability limit such frequency. In the system shown in Figure 3.9, the output of the laser also is passed through a stimulated Raman scattering cell containing high-pressure CH_4. Characteristics of the laser emission at the sample include 20 ps pulse width with a peak flux of approximately 10^9 W/cm^2.

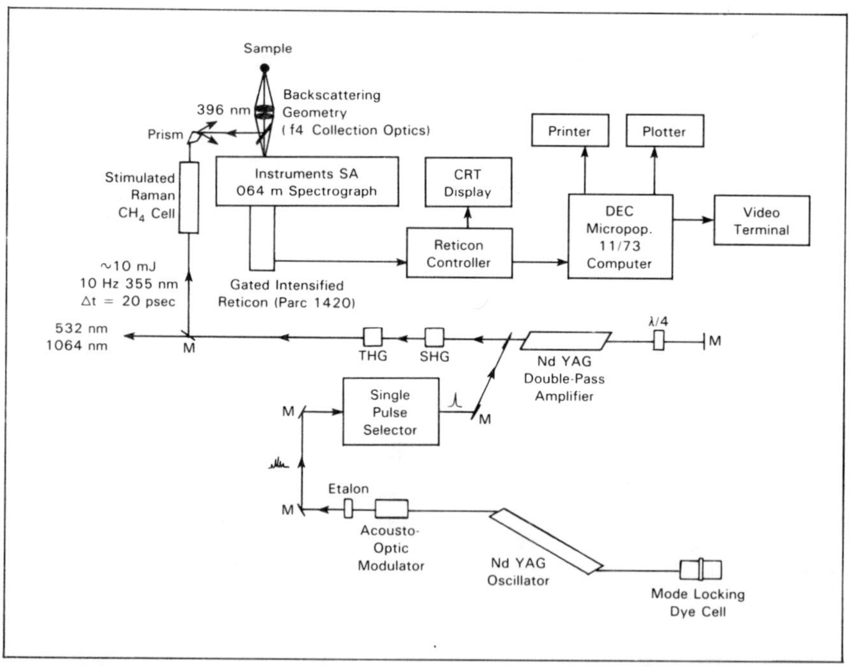

Figure 3.9. Schematic diagram of a picosecond Raman spectrometer (15). Reprinted with permission from the *Journal of Physical Chemistry*. Copyright 1986 American Chemical Society.

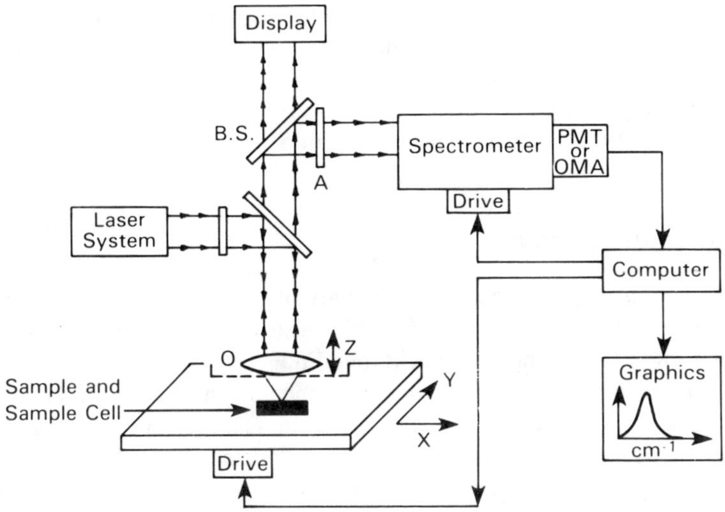

Figure 3.10. Components in a laser microprobe system (18).

Interestingly, most of the components in this picosecond system can be purchased from laser equipment manufacturers.

The use of SERS effects to enhance the observation of molecules at low concentration extends the capabilities of Raman spectroscopy. Hembree, Oswald, and Smyrl (16) described a cell they used for the acquisition of surface-enhanced microspectroscopy of pyridine on Cu, Ni, and Ag surfaces. The diameter of focus of the laser on the sample was less than one micrometer and could be scanned across the surface to map the concentration and association of adsorbed species. Because the SERS effect includes a long-range component, it is not necessary to investigate adsorption directly onto the metals that are known to produce surface enhanced spectra (SES). This has been amply demonstrated by Parry and Dendrums (17), who observed the Raman spectra of polysterene, palmitic acid, and cetyltrimethylammonium bromide at thicknesses as small as 2.0 nm when these thin layers were formed on a 4.0-nm Si substrate, which in turn was located on top of Ag, Au, or Cu substrates. These authors also presented the details of the specialized cell that enabled the acquisition of the Raman data.

Van Duyne, Haller, and Altkorn (18) demonstrated the use of the SERS effect for the observation of extremely low-concentration molecules—10^{-5} mole or 0.17 attomole. The block diagram of a system similar to what they used is shown in Figure 3.10. It is an SERS system coupled with microspectroscopy attachments. The translation stage on which the sample is located can be scanned to bring into Raman excitation selected one-micrometer-diameter areas; these areas can be simultaneously observed by use of a TV monitor. It can be expected with such high-convergence optics in the sampling areas that the laser power at the sample must be limited to less than approximately 50 mW to avoid sample degradation. Such degradation can cause the emission of high-intensity fluorescence, which typically can obscure the observation of normal Raman spectra not containing the 10^6 enhancement found in SERS.

Spectral Enhancement

The use of Raman spectroscopy for the study of catalysts often implies the need to maximize signal intensities from species at low concentrations or spectra of molecular units containing high background noise. As noted above, the Raman effect is an inherently weak scattering phenomenon. In addition, laser excitation in the visible to UV region can cause intense fluorescence or luminescence from samples so that the Raman signal apparently is undetectable. Although electronic transitions responsible for fluorescence generally are understood, the delineation of the origin of such fluorescence from catalysts has been elusive; in fact, such fluorescence should be examined more thoroughly to improve the development of routine Raman measurements on catalysts. However, recently developed spectral and signal enhancement methods have been able to surmount such spectral interference and are extremely important in the application of Raman spectroscopy to catalysts.

As indicated in Chapter 4, the early laser Raman investigations of catalyst structural features showed that the spectral effects of fluorescence could be diminished by "burn-off," which entailed relatively high-intensity laser irradiation of the catalyst for periods of minutes to hours. This continuous illumination has also been described as a photochemical degradation of the interfering fluorophore (19). Such a technique for fluorescence degradation can also degrade the sample. Use of nonlinear spectroscopic techniques, such as coherent anti-stokes Raman spectroscopy (CARS), inverse Raman spectroscopy (IRS), Raman gain spectroscopy (RGS), and Raman induced Kerr-effect spectroscopy (RIKES) can also eliminate unwanted background. However, although some of these nonlinear techniques may be applied to catalysis someday, and are briefly described in Chapter 2, they are neither simple to apply nor convenient for routine measurements.

One technique that has been successfully applied to diminish the effects of fluorescence is pulsed or gated Raman spectroscopy. The success of this technique lies in the fact that Raman scattering is prompt, having lifetimes in the range of 10^{-11} to 10^{-13} second, whereas the lifetime of fluorescence is 10^{-6} to 10^{-9} second (20,21). Hence, a pulsed laser in conjunction with ultrafast gated electronics (as, for example, in Figures 3.8 and 3.9) can be successfully applied to catalysis. The length of the laser pulse has to be shorter than the lifetime of the fluorescent state for pulsed techniques to be effective. In addition, the discrimination against fluorescence is effective only if its lifetime is considerably greater than the response time of the detector used in the experiment, for example, ~0.2 nanosecond for a photomultiplier. A related pulse Raman spectroscopic technique has used the modulated component of emission from a mechanically chopped laser beam and a lock-in detector following the photomultiplier tube (22).

Other methods of decreasing the contribution of fluorescence to Raman spectra of catalysts can be lumped under "differential techniques." Various differential techniques have been applied. For example, frequency modulation, where the laser beam is modulated by a sinusoidal wave, uses the difference between the luminescence and the Raman lifetimes to obtain a difference spectrum, which is the sample spectrum minus the interference of fluorescence (23). Besides sinusoidal modulation, other forms of frequency modulation can be used to provide discrimination; Bright and Hieftje (24) utilized the mode noise contained in the laser signal and a microwave spectrum analyzer, used as a frequency-selective amplifier, to enhance the signal-to-noise ratio in their Raman spectra. Another modulation technique that has been successfully utilized and is related to frequency modulation is called phase-resolved modulation (25,26). In this case, the frequency of modulation is increased to a point where one of the components of the total scattered signal (Raman + fluorescence) undergoes an appreciable phase shift. The difference in phase is detected by a phase-sensitive detector such as a lock-in amplifier.

Wavelength modulation has been employed to decrease the deleterious effects of fluorescence. Here, spectra are acquired by using two different wavelength excitations (27). One of these wavelengths is in the region where the fluorescent response is maximized, whereas the other wavelength excites a combination of fluorescence and the Raman spectrum of the sample. The signals from these spectra are sent to a differential recorder to separate the Raman signal from the background.

Polarization modulation, a differential technique, can be used to enhance spectral quality (28). In this method, the polarization of the Raman bands is used to effectively discriminate between fluorescence and Raman scattering. This discrimination is very effective because fluorescence is not polarized, whereas many bands from solids and liquids are highly polarized. In the case of powdered samples this technique may not be very effective; in the case of adsorption on smooth surfaces it should work very well.

An alternate method to distinguish a weak Raman signal from the background in a sample spectrum has been described by Dyer, Hardin, and Bradley (29). This method entails the use of an adaptive linear predictor in which a low signal-to-noise spectrum is treated by electronically filtering the signal as a combination of weak Lorentzian peaks in a white noise background. A spectrum containing a signal-to-noise ratio as low as 0.3 was shown to be extractable by use of this technique. In some cases, it may be possible to employ a combination of techniques to enhance the quality of Raman data. For example, the use of photoquenching of the species causing the interference in conjunction with an adaptive linear predictor may be expected to enable the acquisition of the spectrum from minute amounts of specific molecular species. Such photoquenching recently has been noted to be more effective for pulsed-laser excitation as compared to continuous excitation (30). In this case, the luminescence background was decreased by factors of 10 to 100 without noticeable deterioration of the sample.

Another method of acquiring Raman spectra in the presence of fluorescence has been described by Mann and Vickers (31). In their approach, the difference in the peak widths of fluorescence bands as compared to Raman bands, where Raman bands are substantially more narrow than fluorescence, was used as a discrimination factor in the Fourier domain after Raman measurements were made on fluorescing samples. Fluorescence generally buries or swamps any underlying Raman signal; such a statement is associated with observation of the output signal of a detector on a Raman spectrometer. However, as pointed out by Mann and Vickers (31), in certain instances measurements can be made on fluorescing samples to assure that the output of the detector is a linear combination of the Raman and fluorescent signals. In this case and for their spectra of a highly fluorescent dye—Erionchrome Blue SE at predetermined solution concentrations of 0 to 50 ppm dye plus HCl in methanol—a linear baseline for individual and weak peaks was employed, in which the large-dc, fluorescent background was neglected. The size of the error caused by using a linear baseline was proportional to the amount of assumed curvature in the baseline and to truncation of the spectra at specific points outside of the spectral peak. Use of a parabolic baseline was also tried and resulted in a concomitant improvement in quantitation of the amount of analyte in solution.

Higuchi, Yu, and Tanaka (32) examined the influence of solvents on the fluorescence background in Raman spectra of polycyclic aromatic compounds. For 1-nitropyrene, it was shown that the fluorescence intensity was dependent on the type of solvent used. All of the polycyclic aromatic compounds studied were insoluble in H_2O; hence, decreasing fluorescence by the method of adding fluorescent quenching agents, such as inorganic salts (33,34), was not possible. The Raman spectra of 1-nitropyrene in 1,1,2,2-tetrachloroethane, acetonitrile, and carbon tetrachloride were extremely different in quality; no discrete Raman scattering could be observed when using 1,1,2,2-tetrachloroethane as solvent. Other than CCl_4, the "good" solvents for obtaining solution spectra were CS_2, C_6H_6, and $CHCl_3$. Further studies are needed to elucidate the influence of solvents on fluorescence from solutes. In particular, studies directed at delineating mechanisms of fluorescence activation and quenching not only would provide fundamental physiochemical information on solvent/solute interactions, but would give experimentalists additional options for Raman data acquisition.

The accuracy of dispersive Raman monochromators can be considered to be approximate ±1 cm^{-1}. It is possible to extend this accuracy a hundred-fold and to measure frequency shifts of Raman bands down to 0.01 cm^{-1}. The technique used to obtain such high accuracy has been called Raman difference spectroscopy (RDS). It generally entails the simultaneous recording of spectra from two or

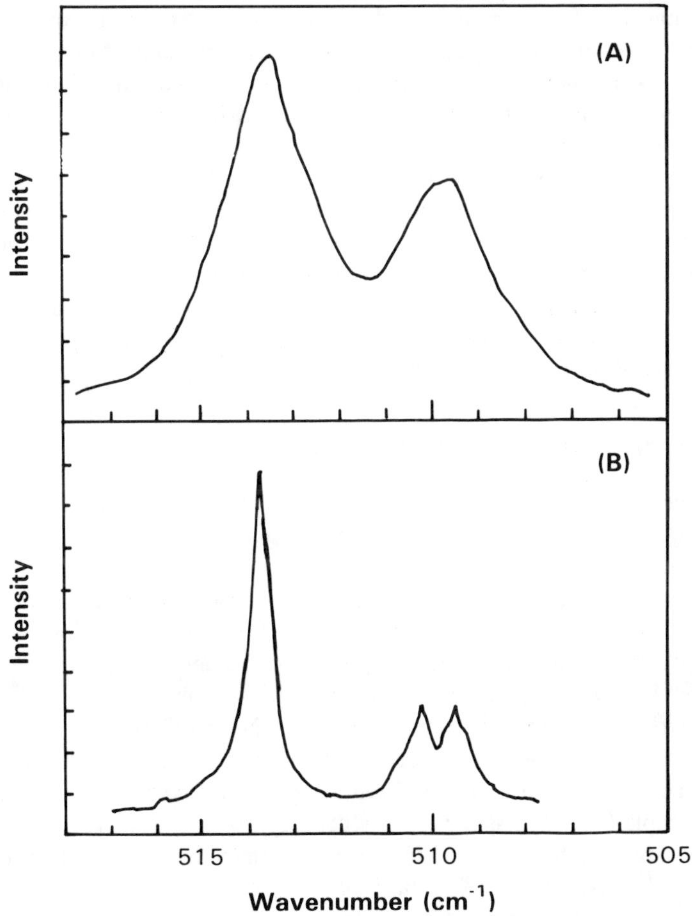

Figure 3.11. Raman spectra of naphthalene at 77°K with laser power of 150 mW at 477 nm. (a) With spectrometer alone, spectral slit width = 1 cm^{-1}, full scale intensity of 660 cps. (b) With interferometer accessory installed with fsr = 6 cm^{-1} and spectrometer spectral slit width = 4 cm^{-1}, full scale intensity of 2200 cps (43).

more samples, with intensity measurements being made separately on each sample. For example, the samples could be alternately positioned in the laser beam for intensity measurements at various frequencies of the spectrum. Such sample positioning can be achieved by interchanging sample cells (35,36) or by using a rotating divided cell containing the two samples in different sectors of the divided cell (36–39). Alternately, the samples can be stationary, and the laser beam can be optically controlled by optical chopping, rotating butterfly mirror, or beam splitter systems (39–42).

Silberman and Springer described the use of a relatively inexpensive interferometric accessory for increasing the resolution of a standard dispersive monochromator (43). They coupled a piezoelectrically driven Fabry-Perot interferometer with a Raman spectrometer and displayed its application in delineating spectral splitting for the 500 to 520 cm^{-1} region of naphthalene at 77°K. Figure 3.11 compares the spectra of this sample with and without the interferometer placed in front of the input optics of the monochromator. Spectral resolution is seen to be enhanced with the interferometer; such enhancement coincided with decreased incident laser power and less amplification of the Raman signal needed to produce signal/noise spectra identical to those obtained without the device. Hence, this relatively simple accessor could be employed to improve detection and analysis of very weak Raman bands.

Interferometric Raman Analyses

The above-described methods for decreasing the influence of or eliminating fluorescence have been successfully applied in the analysis of many types of samples. Fluorescence is primarily attributed to the excitation of impurities and/or other surface species in catalysts that contain absorption bands in the spectral region of laser excitation; so if excitation in other wavelength regions were used, specifically in regions not containing absorption bands, then the fluorescence should no longer be important. Recently this assumption has been proved in experiments employing laser excitation from a Nd-YAG laser at 1064 nm. For example, Figure 3.12 shows the difference in the signal-to-noise that is obtained in Raman experiments by using different wavelengths of excitation of the chromophore sample (44). The spectrum produced by excitation with a Nd-YAG laser has a substantially better signal-to-noise ratio than that produced by the other laser wavelengths, which are in the visible region.

The use of the 1064 nm excitation from a Nd-YAG laser largely avoids the absorption processes that lead to fluorescence. Unfortunately, the change in excitation wavelength from 514.5 nm to 1064 nm decreases by a factor of 16 the Raman scattering cross section because this cross section has a $(\text{wavelength})^4$ dependence (see Chapter 2). Coupled with this decreased sensitivity is the fact that although the detectors that can be used for the near infrared have a higher

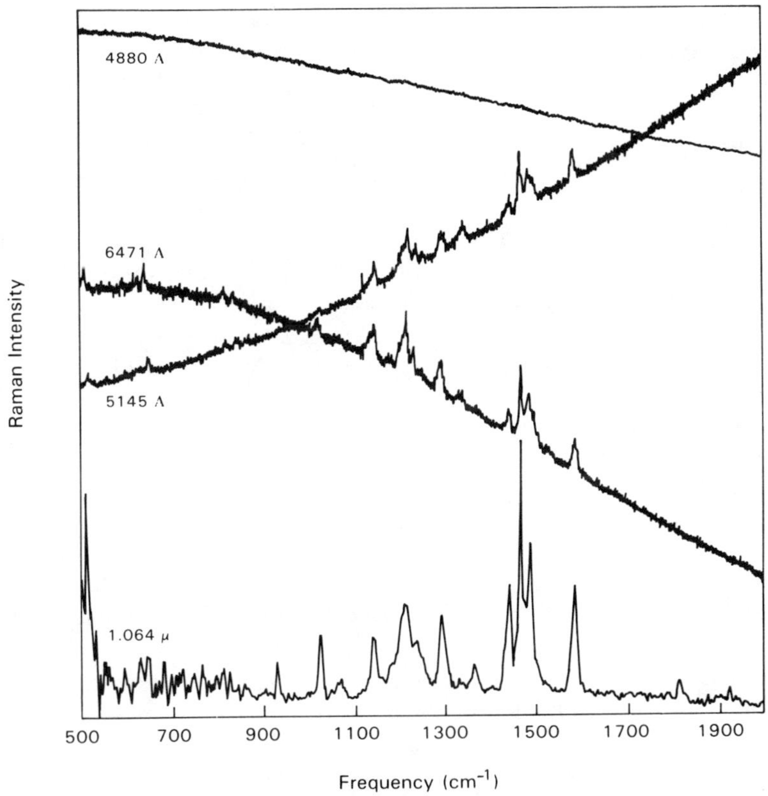

Figure 3.12. Raman spectra of a cyanine dye, 3,3′-dioctadecyl-2,2′-methylene-bisbenzothiazole perchlorate, recorded using different wavelengths throughout the visible and into the near-infrared region. Although attempts were made to keep the resolution constant, the spectrum at 488 nm could not be recorded with slits larger than 2 cm^{-1}. This spectrum as shown was reduced by a factor of four so that it might be directly compared with the other spectra (44).

quantum efficiency (see Figure 3.7), they are also much noisier than photomultipliers. Hence, the successful acquisition of laser Raman spectra from samples using near-infrared excitation has had to rely on the maximization of all experimental and instrumentation parameters to bring its sensitivity in line with what normally can be accomplished with visible excitation.

The diagram in Figure 3.13 shows an interferometric system for Fourier-transform Raman spectroscopy (FTRS), which can be used to obtain good-quality spectra even at 1064 nm excitation. At present, the development and use of FTRS systems can be compared to the conditions of development and use of FTIR systems about 10 or 15 years ago. It is expected that FTRS will become

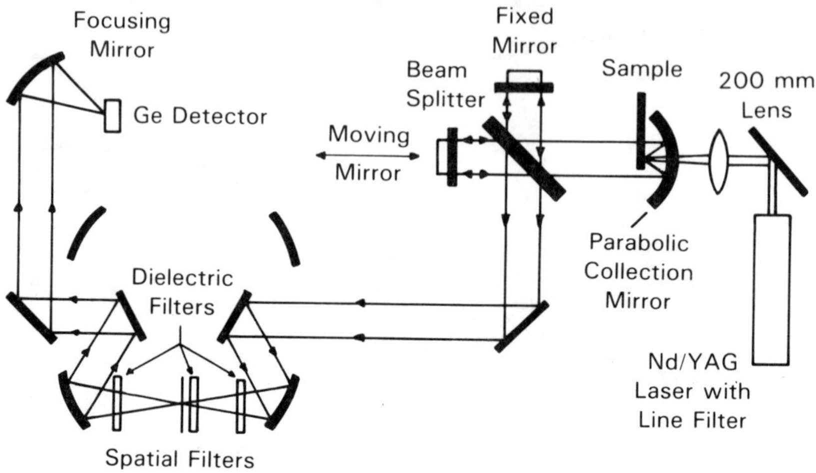

Figure 3.13. Optical components, collection optics, and filtering optics for FT-Raman interferometer to be used for Fourier-Transform Raman spectroscopy.

the workhorse of Raman spectroscopy in the near future, as FTIR has in infrared spectroscopy.

The key to the success of interferometric Raman analysis is effective filtering of the relatively intense Rayleigh scattered radiation. This filtering, or stray light rejection, is on the order of 10^{-3} for a Michelson interferometer, whereas it should be approximately 10^{-10}. Such a high optical density extinction at the laser frequency should also have a very steep transition from complete extinction to complete transmission near the laser frequency. Although such a transmission filter has not yet been developed, the filtering system described by Chase allowed spectral acquisition within 250 cm^{-1} of the 1064 nm excitation from a Nd-YAG laser (45). This system used three dielectric long pass filters that together had a extinction of 10^{10} at the laser line. However, these filters also have secondary extinction regions that, in effect, produced a filter spectrum in spectral regions shifted from the excitation frequency. Removal of some of these filters from the spectrometer, and in effect decreasing the number of extinction bands, was shown to appreciably increase the noise in the FTR spectrum. This broadband noise is, in part, due to the distributive property of the Fourier transform, where noise within one strong band is distributed over an extended spectral region affecting the signal-to-noise ratio in regions far removed from the strong band. Hence, use of the above-described filter system would necessitate storage on a computer of the background transmittance of the filter and subsequent application of this correction each time a spectrum was obtained. Other filtering schemes have been discussed by Hirschfeld and Chase (46). In addition, it is possible to combine a interferometer with dispersive elements to attain the stray light rejection that is needed to perform FTRS.

Although the Raman scattering cross section is decreased by using 1064 nm excitation as compared to the use of visible wavelength excitation, the amount of power that can be incident upon a sample before its degradation is much higher in the near-infrared than in the visible region. Hence, part of the decrease in sensitivity resulting from near-infrared excitation can be accommodated by the higher laser input power. For example, Chase (45) showed that the power at the sample at 514.5 nm excitation could not exceed 150 mW before sample degradation, whereas 1500 mW of power at 1064 nm did not produce any detectable sample degradation.

Accuracy in the measurement of absolute frequency is also better in an FTR spectrometer than in a conventional dispersive instrument, as a result of the laser reference used in FT systems, which provides precision on the order of 0.01 cm^{-1}. Such precision enables the subtraction of spectra without the errors that are inherent in mechanical drives in monochromators. The essential benefits of using an FTR spectrometer as compared to using a dispersive spectrometer have been discussed (44–47). At present there are commercial vendors of FTRS systems; and although experience with these systems is very limited, it is anticipated that developments in system configuration, detectors, and filtering will enable FTRS to become as popular as FTIR.

REFERENCES

1. Kiefer, W., and H. J. Berstein, *Appl. Spectrosc. 25,* 609 (1971).
2. Brown, F. R., L. E. Makovsky, and K. H. Rhee, *J. Catal. 50,* 162 (1977).
3. Cheng, C. P., J. D. Ludowise, and G. L. Schrader, *Appl. Spectrosc. 34,* 146 (1980).
4. Makovsky, L. E., J. R. Diehl, and J. M. Stencel, unpublished design.
5. Stencel, J. M., and E. B. Bradley, *Rev. Sci. Instrum. 49,* 1163 (1978).
6. Arunkumar, K. A., H. A. Marzouk, and E. B. Bradley, *Rev. Sci. Instrum. 55,* 905 (1984).
7. Spohn, P. D., and T. B. Brill, *Appl. Spectrosc. 41,* 1152 (1987).
8. Lin, L.-T., C. K. Mann, and T. J. Vickers, *Appl. Spectrosc. 41,* 422 (1987).
9. Zimmerer, N., and W. Kiefer, *Appl. Spectrosc. 28,* 279 (1974).
10. Kiefer, W., *Adv. Infra. Raman Spectrosc. 3,* 1 (1977).
11. Heritage, J. P., and D. L. Allara, *Chem. Phys. Lett. 74,* 507 (1980).
12. Asher, S. A., C. R. Johnson, and J. Murtaugh, *Rev. Sci. Instrum. 54,* 1658 (1983).
13. Wilke, V., and W. Schmidt, *Appl. Phys. 18,* 177 (1979).
14. Fodor, S. P. A., R. P. Rava, R. A. Copeland, and T. G. Spiro, *J. Raman Spectrosc. 17,* 471 (1986).
15. Rossetti, R., and L. E. Brus, *J. Phys. Chem. 90,* 558 (1986).
16. Hembree, D. M., Jr., J. C. Oswald, and N. M. Smyrl, *Appl. Spectrosc. 41,* 267 (1987).
17. Parry, D. B., and A. L. Dendrums, *Appl. Spectrosc. 40,* 656 (1986).
18. Van Duyne, R. P., K. L. Haller, and R. I. Altkorn, *Chem. Phys. Lett. 126,* 190 (1986).
19. Freeman, S. K., *Applications of Laser Raman Spectroscopy* (Wiley, New York, 1974).
20. Laubereau, A., D. von der Linde, and W. Kaiser, *Phys. Rev. Lett. 28,* 1162 (1972).

21. Lytle, F. E., *Anal. Chem. 46,* 545A (1974).
22. Nemanich, R. J., S. A. Solin, and J. Doehler, *Rev. Sci. Instrum. 47,* 741 (1975).
23. Galeener, F. L., *Chem. Phys. Lett. 48,* 7 (1977).
24. Bright, F. V., and G. M. Hieftje, *Appl. Spectrosc. 40,* 583 (1986).
25. Genack, A. Z., *Anal. Chem. 56,* 2957 (1984).
26. Demas, J. N., and R. A. Keller, *Anal. Chem. 57,* 538 (1985).
27. Morhange, J. F., and C. Hirlimann, *Appl. Opt. 15,* 2969 (1976).
28. Angel, S. M., M. K. De Armond, K. W. Hanck, and D. W. Wertz, *Anal. Chem. 56,* 3000 (1984).
29. Dyer, S. A., D. S. Hardin, and E. B. Bradley, *J. Raman Spectrosc. 15,* 400 (1984).
30. Hamaguchi, H., T. Tahara, and M. Tasumi, *Appl. Spectrosc. 41,* 1265 (1987).
31. Mann, C. K., and T. J. Vickers, *Appl. Spectrosc. 41,* 427 (1987).
32. Higuchi, S., E. J. Yu, and S. Tanaka, *Appl. Spectrosc. 41,* 413 (1987).
33. Furuya, N., A. Matsuyuki, S. Higuchi, and S. Tanaka, *Water Res. 13,* 371 (1979).
34. Furuya, N., A. Matsuyuki, S. Higuchi, and S. Tanaka, *Water Res. 14,* 747 (1980).
35. Amy, J. W., R. W. Chrisman, J. W. Lundeen, T. Y. Ridley, J. C. Sprowles, and R. S. Tobias, *Appl. Spectrosc. 28,* 262 (1974).
36. Chrisman, R. W., J. C. English, and R. S. Tobias, *Appl. Spectrosc. 30,* 168 (1976).
37. Kiefer, W., W. J. Schmid, and J. A. Topp, *Appl. Spectrosc. 29,* 434 (1975).
38. Gardiner, D. J., R. B. Girling, and R. E. Hester, *J. Chem. Soc., Faraday Trans. 71,* 709 (1975).
39. Shelnutt, J. A., *J. Phys. Chem. 87,* 605 (1983).
40. Bodenheimer, J. S., B. J. Berenblut, and G. R. Wilkinson, *Chem. Phys. Lett. 14,* 533 (1972).
41. Martin, J. C., *Rev. Sci. Instrum. 56,* 2217 (1985).
42. Savoie, R., M. Langlais, and R. Beauchesne, *Appl. Spectrosc. 41,* 671 (1987).
43. Silberman, E., and J. Springer, *Appl. Spectrosc. 32,* 352 (1978).
44. Zimba, C. G., V. M. Hallmark, J. D. Swalen, and J. F. Rabolt, *Appl. Spectrosc. 41,* 721 (1987).
45. Chase, D. Bruce, *J. Am. Chem. Soc. 108,* 7485 (1986).
46. Hirschfeld, T., and B. Chase, *Appl. Spectrosc. 40,* 133 (1986).
47. Moffatt, D. J., H. Buijs, and W. F. Murphy, *Appl. Spectrosc. 40,* 1079 (1986).

4
Surface and Bulk Structural Chemical Analyses of Catalysts

INTRODUCTION

Raman spectroscopy was not believed applicable to the study of the chemistry and structure of surfaces and/or surface-supported species less than 15 years ago. This misconception was a result of the inherently low intensity of Raman scattered radiation and the fact that the use of lasers as excitation sources was still in its infancy. It was well known that gases, liquids, and solids, both crystalline and noncrystalline, could be studied quite easily with normal Raman (NR) techniques if geometrical and sampling considerations were optimized. However, these successes in NR spectroscopy appeared characteristic of "bulk" investigations; the surfaces of materials were not considered. It is now well known that NR spectroscopy can provide detailed information on an extremely wide variety of surfaces and surface-supported phases. That this can be accomplished is probably somewhat surprising, albeit exciting, to molecular spectroscopists who were taught the fundamentals of molecular spectroscopy in the 1950s to early 1970s. An entirely new set of theoretical and experimental considerations has been sought and delineated within the last 10 to 15 years to accommodate surface Raman spectroscopy. Such development is continuing at a rapid pace with the emergence of research groups that are starting to employ normal Raman spectroscopy to address questions of catalyst speciation.

This chapter presents a comprehensive survey of NR investigations published in recent years, especially 1977 through 1988. The types of catalysts studied, the rationale for defining the surface species that are present, and the importance of the vibrational information to catalysts are considered. A summary of the types of catalysts examined by Raman spectroscopic techniques is presented in Table 4.1. Structural and chemical information obtained by use of Raman spectroscopy on these supported catalysts is presented in the following section ("Catalyst Structural Determinations"), whereas "Raman Studies of Adsorbed Species"

Table 4.1. Catalysts and catalyst surfaces studied by Raman spectroscopy.

TYPE	FORM	USE
Li, Na, K	Metal/oxide	Promoters
Fe	Metal	NH_3 synthesis, Fischer-Tropsch (FT)
Co	Metal	Hydrogenation, FT, oxidation, promoter
Ni	Metal	Hydrogenation, dehydrogenation, oxidation, promoter
Cu	Metal/oxide	Methanol synthesis, electrocatalysis, FT promoter
Rh	Metal/oxide	Hydrogenation
Pd	Metal/oxide	Methanol synthesis
Ag	Metal	Electrocatalysis
Cd	Metal	Electrocatalysis
In	Metal	Electrocatalysis
Sn	Metal/oxide	Reforming
Os	Metal/oxide	Isomerization
Ir	Metal/oxide	Reforming
Pt	Metal/oxide	Reforming, electrocatalysis
Au	Metal/oxide	Electrocatalysis
MgO	Metal/oxide	Scavenger, support
Al_2O_3	Oxide	Support, sorbent, olefin isomerization
SiO_2	Oxide	Support, sorbent
CaO	Oxide	Scavenger, support
TiO_2	Oxide	Support, alcohol dehydration
V_2O_5	Oxide	Oxidation and (catalytic poison)
Cr_2O_3	Oxide	Dehydrogenation, oxidation, promoter
Mn_3O_4	Oxide	FT promoter
ZnO	Oxide	Methanol synthesis, isomerization, hydrogenation
ZrO_2	Oxide	Alcohol dehydration
MoO_3	Oxide/sulfide	HDS, metathesis
La_2O_3	Oxide	Alcohol synthesis
HfO_2	Oxide	Alcohol dehydration
WO_3	Oxide/sulfide	HDS, metathesis
Re_2O_7	Oxide	Metathesis
Bi_2O_3	Oxide	Oxidation
Zeolites A, X, Y, ZSM-5	Oxide	Cracking
Molybdates/iso-polymolybdates	Oxides	Photocatalysis, oxidation
Montmorillonite	Oxides	Cracking

discusses Raman data on molecular species adsorbed on high-surface supports, smooth metal surfaces, and single metal crystals. The final section ("Zeolites—Synthesis, Structure, and Deactivation") presents recent Raman studies on zeolites, including their synthesis, delineation of structural characteristics, and deactivation by carbonaceous and metal poisons.

CATALYST STRUCTURAL DETERMINATIONS

Part of the impetus to use laser Raman spectroscopy for eludidating the chemical and physical properties of supported catalytic phases came from the Raman data first published by Brown, Makovsky, and Rhee (1–3). Prior to these publications, Raman spectroscopy of catalysts had focused predominantly on the characteristics of supports and/or the adsorption of probe gaseous molecules such as pyridine and others (see section on "Raman Studies," below). Villa, Trifiro, and Pasquon (4) in 1974 had discussed the inability of Raman spectroscopy to detect Mo-associated species on gamma-Al_2O_3, concluding that with ~10% MoO_3/Al_2O_3 a new MoO_3 compound was formed that did not have Raman active modes. In addition, Beuchler and Turkevich (5) in 1972 had published Raman results on Vycor glass and MoO_3/Vycor. However, these spectroscopic data concentrated on the relatively weak scattering from Vycor itself, and not on the supported phase. In fact, the relatively low intensity of Raman spectra from amorphous supports such as SiO_2, Al_2O_3, and SiO_2-Al_2O_3 is advantageous for the acquisition of good-quality spectra of supported phases. Since the reports of Brown, Makovsky, and Rhee, laser Raman spectroscopy has been used extensively to provide otherwise unobtainable information on supported catalytic phases. This chapter describes and compares these efforts.

Supported MoO_3

MoO_3/Al_2O_3

The initially reported work of Brown et al. used commercially prepared MoO_3/gamma-Al_2O_3 and CoO-MoO_3/Al_2O_3 catalysts (1,2). These MoO_3/Al_2O_3 catalysts were of interest because of their application in metathesis, oxidation, and hydrogenation of hydrocarbons, water gas shift, CO hydrogenation, and especially the hydrodesulfurization and hydrodenitrogenation of petroleum products and coal. Bands associated with Mo═O stretching modes, expected to be located in the 800 to 1000 cm^{-1} region (6), were observed in the commercial, pelleted catalysts. Upon sulfidation, these bands disappeared with the concomitant appearance of bands due to Mo–S stretch at 405 and 372 cm^{-1}. Interestingly, used/sulfided pelleted catalysts were also examined. These spectra contained intense bands due to C–C stretching vibrations similar in position and relative intensity to those described for carbon deposits under "Zeolites," below.

The Raman spectra from commercial MoO_3/Al_2O_3, NiO-MoO_3/Al_2O_3, and CoO-MoO_3/Al_2O_3 pellets and extrudates and the effects of the concentration of Ni, Co, and MoO_3, the calcination procedure, and the addition of H_2O/D_2O on the types and genesis of surface species were discussed (2). Bulk, crystalline MoO_3 was detected only after molybdena loadings reached a theoretical monolayer level, assuming a 0.15 nm^2 size for the supported molybdate. Formation of crystalline MoO_3, observed with 20% molybdena for ~200 m^2/g Al_2O_3 and with 10% molybdena for ~100 m^2/g Al_2O_3, was tentatively related to a corresponding decrease in catalytic dehydrogenation activity (7).

Some of the previously held theories on the surface structure of supported MoO_3, in which the molybdena was thought to consist of a monolayer that retained the properties of bulk MoO_3 (8–10), were subsequently reexamined as a result of the publications of Brown et al. (1–3). For example, the Raman results indicated the presence of both tetrahedral and octahedral Mo species. There was general disagreement in the literature at that time about the coordination of Mo on alumina surfaces, which had been discussed as primarily tetrahedral (11), primarily octahedral (9), or a combination of both (12,13). However, the use of commercial catalysts as a basis for the Raman data probably limited somewhat the scope of the research because fundamental parameters in catalyst synthesis, such as impregnating pH, surface area of the support, calcination conditions, and so forth, could not be examined systematically.

The concentration of MoO_3 was varied, and spectra from external and internal surfaces of these MoO_3/Al_2O_3 pellets were acquired (see Figure 4.1) (2,13,14). Segregation of various Mo-containing species was noted. Such segregation could have been caused by various factors: (1) There could have been a weak interaction of molybdate ions with the support, which during the penetration of the impregnating solution into a pellet caused a relative decrease in the molybdenum concentration toward the interior of the pellet. Here, a slower diffusion of molybdate ions through water-filled pores would also enhance Mo concentrations on the external layer of the pellet. (2) Even if molybdenum was uniformly distributed in the pellet, drying of the impregnated pellets could have carried H_2O out over deposited molybdena, possibly redissolving it. These ions in solution could then be deposited on the external portion of the pellets. The development of specific Raman bands in Figure 4.1 as a function of Mo loading with a 212 m^2/g surface area support was similar to the development of Raman bands for MoO_2/Al_2O_3 with 110 m^2/g surface area support. In the latter case, the appearance of Raman bands occurred at approximately one-half the loading at which they had appeared in the former catalysts. The bands in Figure 4.1 were assigned by comparison to spectra of known compounds such as bulk, crystalline MoO_3 and $MoO_3 \cdot 2H_2O$ (14–19).

X-ray diffraction data of crystalline MoO_3 indicate that there are three distinct molybdenum–oxygen structures, having: (1) one terminal oxygen bound to only one Mo atom, (2) two bridged oxygens bound to two Mo atoms, and (3) three bridged oxygens bound to three Mo atoms. Hence, from Figure 4.2, the 998 cm^{-1} band for crystalline MoO_3 would be due to terminal Mo═O stretching, the 821 cm^{-1} band a result of Mo–O–Mo stretching, and the 668 cm^{-1} band a result of the three-bridged oxygen stretching motion. Intensity differences in a

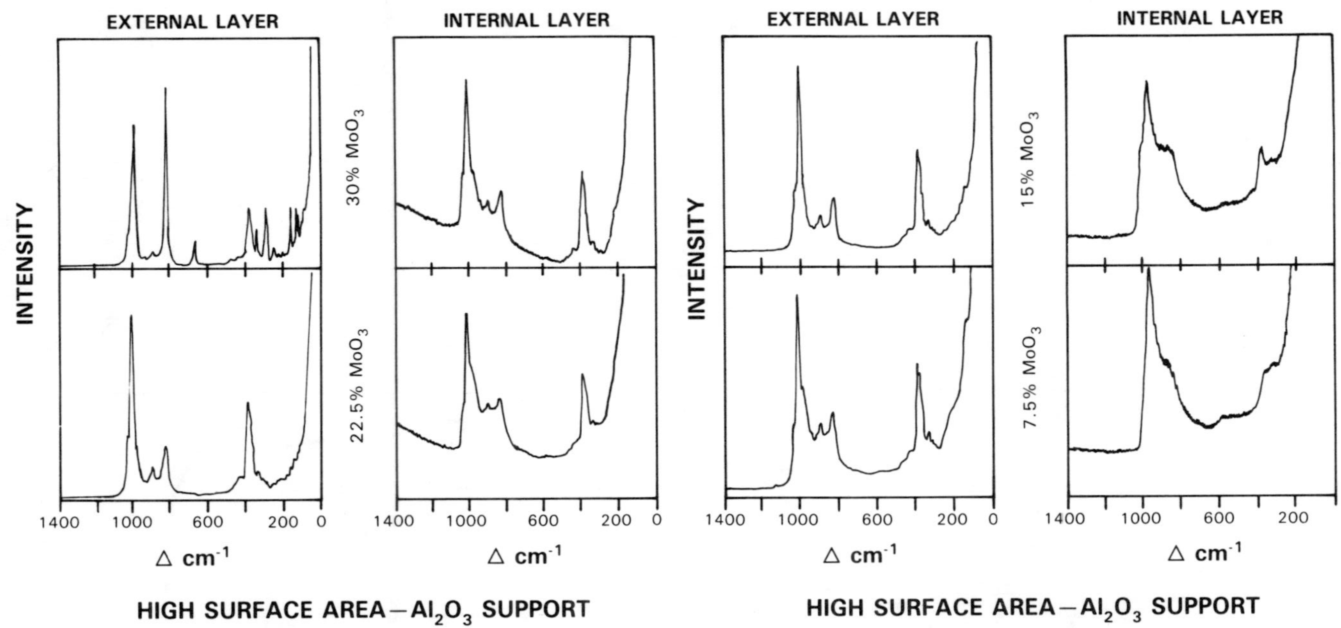

Figure 4.1. Raman spectra of the external and internal layers of molybdenum-on-alumina (~200 m^2/g surface area) catalysts. Spectra of the external layers were obtained first, the catalyst pellet was machined to remove a thin cylindrical layer, and then spectra of internal layers were obtained (14).

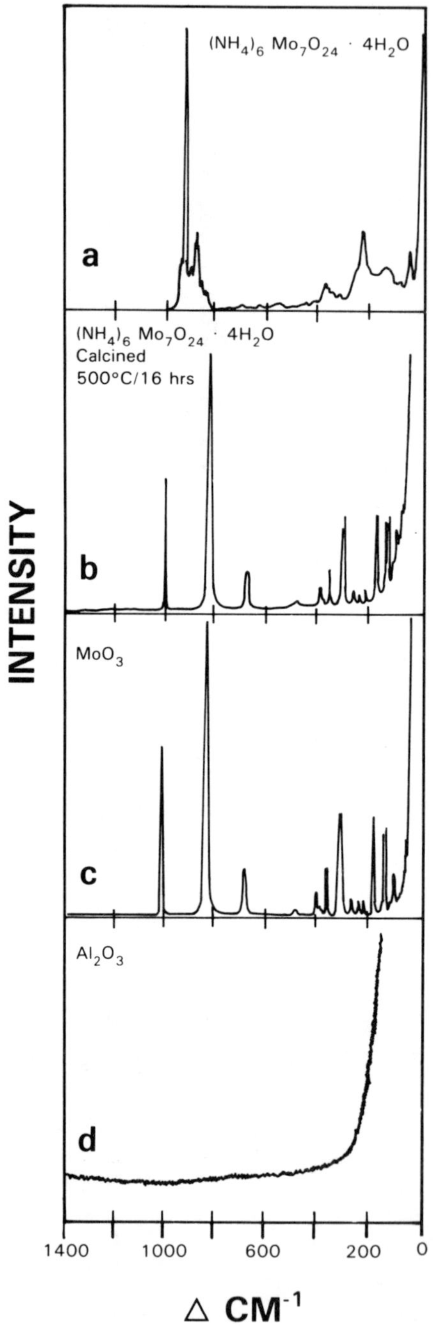

Figure 4.2. Raman spectra of: (a) ammonium paramolybdate; (b) ammonium paramolybdate calcined in air at 500°C for 16 hours; (c) molybdic anhydride; (d) alumina support.

821 cm^{-1} and a 1006 cm^{-1} band from such catalysts (see Figure 4.1), as compared to the 821 and 996 cm^{-1} bands of crystalline MoO_3 in Figure 4.2, suggested that the Mo-surface species contained terminal/double-bridged oxygen stoichiometry substantially different from MoO_3—and the absence of a band at 668 cm^{-1} in the catalysts also suggested that no triply bridged oxygen species were present in the supported phase.

The dependence of the formation of crystalline MoO_3 on the Mo loading observed by Brown et al. (1–3) was also observed in laser Raman spectra obtained by Medema et al. (15). Here, the Raman spectra suggested the presence of isolated, tetrahedral Mo species at loadings of 5% MoO_3, octahedrally coordinated species at loadings of 10 to 15% MoO_3, and the subsequent formation of subsurface $Al_2(MoO_4)_3$ and surface, crystalline MoO_3 at loadings near 20 to 30% MoO_3. The genesis of these species and their relationship to the pH of the impregnating solution was discussed by Jeziorowski and Knozinger (20). As pointed out, the existence of polymolybdate anions—Mo_7O_{24}—having octahedral Mo, or tetrahedral Mo, as in MoO_4^{2-}, is controlled by the pH of the solution and any changes in pH upon contact of the solution with an acidic surface such as Al_2O_3. The equilibrium of these species is (21–23):

$$7MoO_4^{2-} + 8H^+ \rightleftarrows Mo_7O_{24}^{6-} + 4H_2O$$

$$(pH = 11) \qquad (pH = 6)$$

Raman spectra of these polyanions and of aqueous Mo_6O_9- and Mo_8O_{26}-ions have been described (21–25). Since the coordination of MoO_3 surface species had been observed to vary as a function of MoO_3 loading while the corresponding condensed nature of the species also varied, it was natural to compare spectra of the surface species with the spectra of polyanions. These polyanions in solution contain characteristic band positions; for example, terminal Mo=O stretching at 943 cm^{-1} for $Mo_7O_{24}^{6-}$ and at 965 cm^{-1} for $Mo_8O_{26}^{4-}$, indicating an increase in frequency with increased degree of polymerization. However, upon calcination of the catalysts and spectral acquisition under ambient conditions, the spectra exhibited no significant differences in Mo=O stretching frequencies at low constant MoO_3 loadings. For increasing Mo content, the Mo=O stretching frequency did increase when the Raman data were collected with the catalysts exposed to ambient conditions.

The formation of a strong band in the 940 to 980 cm^{-1} region, independent of impregnating solution (20), was essentially in agreement with the results of Brown et al. (1–3). Even so, the frequency ranges of generalized Mo—O stretching modes in polyanions, for example, 200 cm^{-1} for Mo—O—Mo deformation, 310 to 370 cm^{-1} for M=O bending, 500 to 650 cm^{-1} for symmetric

Mo—O—Mo deformations, 700 to 850 cm^{-1} for antisymmetric Mo—O—Mo stretching, and 900 to 1000 cm^{-1} for the symmetric and antisymmetric Mo═O stretches, were in general agreement with the frequency ranges in which bands were observed in the catalysts. This fact allowed insight into the possible coordination and the condensed nature of the surface species, as, for example, with 3% MoO_3/Al_2O_3 a band near 200 cm^{-1} was formed upon calcination of the catalyst, thereby suggesting the formation of Mo—O—Mo bonding. In addition, the position of the Mo═O terminal stretching for 8% MoO_3 (pH 11 impregnation) was at 961 cm^{-1}, whereas for a pH 6 impregnated catalyst the corresponding position was at 950 cm^{-1}. Although this shift may imply a change in the condensed nature of the surface species, other complications in Raman spectral acquisition subsequently were proved to exist.

The band associated with the molybdena surface species, a Mo-containing species having relatively high dispersion and interacting with the support, was located near 950 cm^{-1} in all of the above-described data. A subsequent study that used Raman spectroscopy with X-ray photoelectron spectroscopy (XPS), ion scattering spectrometry (ISS), and X-ray diffraction provided a comprehensive picture of the genesis of Mo speciation as a function of Mo loading on ~200 m^2/g gamma-Al_2O_3 (26). It was shown that the Mo was primarily tetrahedrally coordinated at less than about 1/3 theoretical monolayer loadings, whereas the concentration of octahedrally coordinated Mo increased with increasing Mo concentrations up to about a monolayer capacity. This speciation is indicated in Figure 4.3. The generation of crystalline MoO_3 and $Al_2(MoO_4)_3$ was in agreement with previous studies (1–3,14–16), whereas the percentage of tetrahedral Mo (~25%) and octahedral Mo (~65%) at the monolayer loading of 20% MoO_3 was also in agreement with other studies on MoO_3/Al_2O_3 catalysts that had shown increased octahedral Mo with increasing Mo loadings (27,28). The preponderance of tetrahedral Mo of an irreducible nature at less than 5% MoO_3 loadings suggested the possible catalytically unreactive nature of this species (29). In addition, it was shown (30) that crystalline MoO_3 could be redispersed on the support surface without its loss from the surface; extended calcination times of the samples having less than a theoretical monolayer concentration but also having crystalline MoO_3 caused migration of Mo into subsurface regions with the formation of tetrahedrally coordinated Mo as in $Al_2(MoO_4)_3$. Coordination of the highly dispersed Mo surface species, originally termed the "interaction species" (1–3) and observed in most Raman studies of MoO_3/Al_2O_3 catalysts, was not directly discernible by characteristic bands in the Raman spectra.

The Raman results of Jeziorowski and Knozinger (20) suggested that the adsorption of anions from the impregnating solution occured via MoO_4^{2-} independent of the initial pH of the solution. However, it was suggested that the actual pH in the filled pores of the support was greater than 8.5 even at an initial pH of 6 because of the buffer capacity of the Al_2O_3 (32). To avoid complications

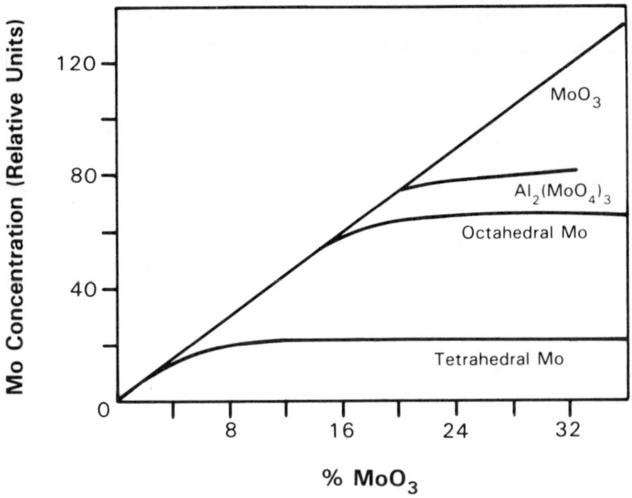

Figure 4.3. Molybdenum speciation as a function of MoO_3 content for a Mo/γ-Al_2O_3 catalyst; catalyst calcined at 500°C for 16 hours (26). Reprinted with permission from *The Journal of Physical Chemistry,* Copyright 1980 American Chemical Society.

associated with buffering capacity, catalyst preparation techniques were used that included the adsorption of molybdate ions from dilute solutions at pH 0.8, 2.0, and 8.6. The final pH was also monitored to ensure that any change in pH upon interaction of the solution with the support was small and would not change the type of solute species; thus the respective final pH values were 1.0, 3.9, and 8.6. The resulting Raman spectra of these catalysts, the band frequencies of which are listed in Table 4.2, were interpreted in terms of aggregated polymolybdates having increased polymerization as a function of increasing loading. In fact, the degree to which the coordination of Mo changed from tetrahedral to octahedral was suggested to be measured by obtaining the intensity ratio of Mo═O bands at 950 cm^{-1} and 970 cm^{-1}. The positions of low frequency bands in Table 4.2 were in general agreement with previous Raman data, and, in comparison with previous data (20), showed an influence of the support in modifying the pH of the impregnating solution and in possibly determining the type of molybdate species that was adsorbed. As was first discussed by Sonnemans and Mars (10), adsorption of anions from aqueous solution is possible when the pH of the solution is adjusted below the isoelectric point (point of zero charge) of the adsorbent; for Al_2O_3 the isoelectric point is near 8.4 (33). Discussion of the effect of the isoelectric point on adsorption while using aqueous impregnation techniques also can be found in references 10, 33, and 34.

Definition of the concentration of octahedral and tetrahedral Mo by monitoring

Table 4.2. Positions (cm^{-1}) of Raman bands for molybdena/alumina catalysts prepared as a function of pH of impregnating solution (32).

pH 8.6 (1.0 atom/g × 10^{-20})				pH 3.9 (4.0 atom/g × 10^{-20})			
SOLUTION	WET	DRIED	CALCINED	SOLUTION	WET	DRIED	CALCINED
				215	217	216	215
318	326	326					
				365	364	365	365
	547	545					
846							
							860
896	905	908		895	906	905	
					928	928	
	945	945	950	940	951	950	950
			965				970

pH 1.0 (7.4 atom/g × 10^{-20})			
SOLUTION	WET	DRIED	CALCINED
215	218	216	215
368	366	365	365
			860
905	903	895	
	930	930	
965	956	955	950
980			975

changes in the position of the ca. 950 cm^{-1} Mo=O stretching band would be an exciting development in the definition of Raman spectra from molybdena-containing catalysts. The data of Cheng and Schrader (31) and other studies (1–3,14–16), however, did not suggest any systematic shift in Mo=O stretching band positions with increased Mo loading or any possible definition of tetrahedral- and octahedral-sensitive Raman bands. In fact, the spectra of MoO_3/Al_2O_3 prepared by incipient wetness, equilibrium adsorption, and decomposition of $Mo(CO)_6$ were almost identical (35), if the presence or absence of crystalline MoO_3 was taken into account. Such development of characteristic molybdena surface species independent of the impregnation technique implies that Mo anions are very mobile once they are on the Al_2O_3 surface, and that the properties of the Al_2O_3 surface to a great degree control the Mo speciation subsequent to calcination of the impregnated catalyst.

Thomas, de Beer, and Moulijn (36) showed that the surface molybdena phase changed with the concentration of the impregnated MoO_3. The catalysts studied,

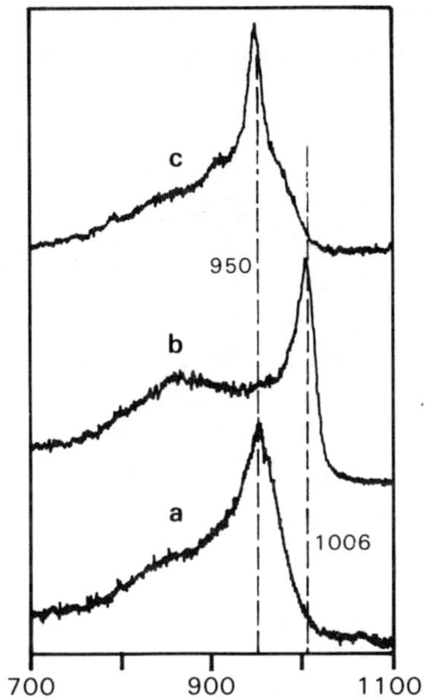

Figure 4.4. Raman spectra of 15% MoO_3/Al_2O_3 for (a) the catalyst exposed to atmospheric conditions for three months, (b) the catalyst calcined in 760 Torr O_2 at 550°C for one hour with O_2 in the sample cell during spectral acquisition, and (c) the catalyst after exposure to 11 Torr of H_2O vapor at 25°C for one hour subsequent to O_2 calcination (38).

from 0.5% to 18.6% (0.09 Mo/nm^2 to 4.48 Mo/nm^2), exhibited temperature programmed reduction (TPR) curves and activity for hydrodesulfurization of thiophene that were a function of the MoO_3 concentration. At higher molybdena concentrations, the surface MoO_3 phase was suggested to contain two- and/or three-dimensional species, the exact description of which was not elucidated. Of course, the application of these catalysts depends on their use in the sulfided phase. Sulfidation causes the catalysts to become black, a less than ideal circumstance for acquisition of good-quality Raman spectra using visible wavelength laser excitation. However, Schrader and Cheng (37) were able to describe from laser Raman spectra of sulfided MoO_3/Al_2O_3 a model for the surface molybdenum that included oxysulfide species, terminal and bridging Mo—S bonding, and anion vacancies.

A cause of the change in the terminal Mo═O stretching frequency as a function of MoO_3 concentration was elucidated by Stencel et al. (38). Figure 4.4 shows some of their Raman data for spectra obtained from in situ O_2-calcined and subsequent H_2O-exposed samples. When the samples were studied under ambient conditions, there was a continuous increase in the M═O stretching frequency with increasing Mo concentration. However, when the samples were kept under controlled environments after calcination, the M═O stretching frequency was nearly independent of the Mo concentration. Instead, the M═O terminal stretching frequency consisted of a band at ca. 1006 cm^{-1} for loadings above 10% MoO_3 and 985 cm^{-1} for loadings below 7% MoO_3. Subsequent exposure of the catalysts to H_2O or D_2O produced spectra that were similar to the originally air-exposed spectra. Hence, adsorption of H_2O was suggested to possibly change the coordination of the surface molybdena, which in turn would cause changes in the M═O stretching frequency. This effect of H_2O was subsequently described for MoO_3, WO_3, and V_2O_5 on TiO_2 and Al_2O_3 supports (see below). The original Raman results of Brown, Makovsky, and Rhee (3) had described some of the same phenomena noted by Stencel et al. (38).

Further data on the surface speciation of molybdena on MoO_3/Al_2O_3 catalysts were presented by Leyrer et al. (39). Here, the appearance of bands in the 220 cm^{-1} range was discussed as characteristic of the Mo—O—Mo deformation of a surface polymolybdate, while a low frequency component of a two-component band in the 300 to 400 cm^{-1} region was considered to be the result of tetrahedrally coordinated Mo. A compilation of bands associated with MoO_3/Al_2O_3 and standard reference compounds is given in Table 4.3. These data complement those that appeared in a recent review of some of the Raman spectroscopic research that has been performed on catalysts (40).

The method of impregnating molybdenum does not radically change the sur-

Table 4.3. Raman bands associated with MoO_3/Al_2O_3 and with reference compounds.

SAMPLE	FREQUENCY (CM^{-1})							
Hydrated MoO_3/Al_2O_3								
3% MoO_3		325				938		
5% MoO_3		330	360		869	943		
10% MoO_3	210	330	360		866	954		
15% MoO_3	208		365		862	959		
20% MoO_3	205,215		365		861	963		
25% MoO_3	204	288	377		820,892,	968,997,1005,1025		
Dehydrated MoO_3/Al_2O_3								
5% MoO_3					870	985		
15% MoO_3					860	1006		
Reference Compounds								
Crystalline MoO_3			668	821			998	
Crystalline $Al_2(MoO_4)_3$			390	820				1006

face speciation of the molybdena on Al_2O_3 when the catalysts are in the calcined state and over a range of concentrations that are less than approximately a theoretical monolayer level (20,32,35,36,38,39). This fact was demonstrated when crystalline MoO_3 was mechanically mixed with a number of supports, Al_2O_3 to be noted, and the resulting mixture calcined and analyzed with Raman spectroscopy (41). Even though the mixture before calcination contained intense peaks characteristic of crystalline MoO_3, after calcination at 720°K the surface of the Al_2O_3 was covered with a molybdena species having spectra characteristics similar to those obtained with other preparation methods. Of course, the concentration of molybdena at which crystalline MoO_3 forms and the physical characteristics of the catalyst will be affected by the preparation procedure; subsequently, the catalytic activity may also be affected. Definitely, the isoelectric point of the support has to be considered. Whether MoO_4^{2-} or $Mo_7O_{24}^{6-}$ is adsorbed directly onto an Al_2O_3 support is still under debate. A recent Mo-95 NMR study (42) has suggested that, upon comparison of the Mo-95 NMR signal from Mo solutions having different pH, impregnation at high pH levels produces predominantly MoO_4^{2-} species in the pores of the Al_2O_3. Kasztelan et al. (43) also showed that high pH levels during impregnation produced a surface molybdena that was easily removed by a water wash; on the other hand, low-pH impregnation produced a molybdena that was not removed by a water wash. Hence, a MoO_4^{2-} species is expected to be weakly bound to the Al_2O_3 surface.

The Raman spectra of reduced MoO_3/Al_2O_3 catalysts were obtained by Payen et al. (44). The data showed the stepwise reduction of Mo(VI) to Mo(V) species, represented by Mo(V)—OH, and subsequent production of Mo(IV), represented by Mo(x)—O—Mo(IV), where x = V or IV species. Reduction of the molybdena at low MoO_3 loadings was found to be difficult, in agreement with previous reduction experiments (26). Reoxidation of reduced catalysts restored the Raman spectra to where they were before reduction, thereby indicating that Mo—O—Al bonding was not irrevocably destroyed by reduction.

Shimada et al. (45) examined the effects of various supports on the catalytic activity and surface properties of molybdena-containing catalysts, and found that for hydrogenation of 1-methylnaphthalene the order of activity for these supports was $Al_2O_3 > TiO_2 > MgO > SiO_2$, whereas for hydrocracking of diphenylmethane and hydrodesulfurization of dibenzothiophene the order of activity was $TiO_2 > SiO_2 > Al_2O_2 > MgO$. Unfortunately, examination of these supported MoO_3 catalysts by Raman spectroscopy was not done under controlled environments, except for the sulfided catalysts. In the oxide form of the catalysts, Shimada et al. observed the characteristically broad Mo═O stretching bands in the 900 to 1000 cm^{-1} region, and, in the case of TiO_2 and SiO_2 supports, crystalline MoO_3 also was monitored. A plot of the Mo═O stretching frequency as a function of Mo loading showed increasing values from 950 to 965 cm^{-1} with concentrations in the range of 0 to 8 atoms/nm^2. This type of shift was interpreted in terms of an increasingly aggregated Mo surface species, but such an interpretation previously had been shown to be questionable because adsorption of H_2O from the atmosphere produces the same type of shifts as a function of Mo loading (38).

Measurement of quantitative Raman spectra from catalysts in most cases would be extremely helpful, to determine the importance of Raman-detected species in catalytic reactions and to compare these molecular results with quantitative elemental results from X-ray photoelectron spectroscopy. Such quantitation was performed for MoO_3/Al_2O_3, and other Al_2O_3 supported catalysts (46). The technique used for such quantitation was to mix KNO_3 of known weight with various concentrations of molybdena or tungsta on Al_2O_3. The integrated intensity of the symmetric stretching mode of NO_3 at 1050 cm^{-1} was then compared to the intensity of the Mo—O—Mo stretching vibration at 820 cm^{-1} and the WO_3 stretching vibration at 808 cm^{-1}. The accuracy of this method was determined to be $\pm 6\%$ at concentrations of MoO_3 and WO_3 as low as 0.1%. In addition, Raman scattering cross sections of crystalline phases of WO_3 or MoO_3 as compared to the cross sections of the respective highly dispersed surface species were determined accurately by use of this internal standard method. For example, the relative cross section of crystalline MoO_3 to surface molybdena was 16 : 1, and the relative cross section of crystalline WO_3 to surface tungsta was 137 : 1 (46). Such an internal standard method is expected to be a preferred method of quantitation, when experimentally possible, as the standard and species of interest are measured simultaneously from one sample. Possible errors in positioning of samples and slight differences in optical alignment or in laser power do not hamper the application of this method.

MoO_3/SiO_2

Fewer Raman studies have been performed on MoO_3/SiO_2 catalysts than on MoO_3/Al_2O_3 catalysts. Part of the reason for this difference is a decreased interest in these catalysts for industrial use; also molybdena dispersion is generally less for SiO_2 supports as compared to Al_2O_3 supports—that is, crystalline MoO_3 will form at much lower Mo loadings on SiO_2 than on Al_2O_3. The isoelectric point for SiO_2 supports is near a pH of 2; molybdena–SiO_2 interactions under other conditions of catalyst preparation will be weak, although a recent study has indicated that it is possible to extend the pH over an broad range (47). Hence, only crystalline MoO_3 was observed on SiO_2 by Jeziorowski et al. (48). The same speciation was noted by Leyree et al. (39), with only very weak Raman intensities detected that could be attributed to Mo–SiO_2 surface interactions. It must be noted that the sensitivity for detection of crystalline MoO_3 is 17 times more than that for detection of surface molybdena (46). Thus, the amount of detected surface molybdena apparently will be small in any Raman spectrum also containing crystalline MoO_3.

Recent studies of MoO_3/SiO_2 that included the use of Raman and X-ray photoelectron spectroscopy and ion scattering spectrometry investigated the effects of hydration–dehydration of the catalyst (49,50). The Raman spectra of 8% MoO_3 after air stabilization and subsequent in situ calcination and H_2O exposure are shown in Figure 4.5. In an air-stabilized state, the major Raman bands were located at 997, 820, and 290 cm^{-1}, having positions and relative intensity in agreement with those for crystalline MoO_3. A relatively weak band also was observed at 950 cm^{-1}. Upon O_2 calcination, the most intense bands

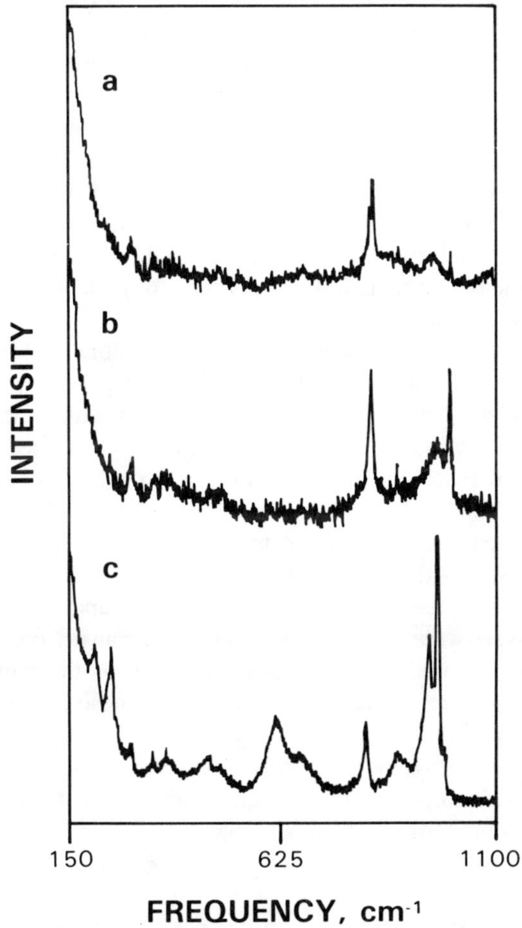

Figure 4.5. Raman spectra of impregnated 8% Mo/SiO_2 catalyst: (a) air-stabilized; (b) in situ O_2-calcined at 550°C; and (c) H_2O-exposed (49). Reprinted with permission from *The Journal of Physical Chemistry,* Copyright 1986, The American Chemical Society.

were located at 820, 950, and 997 cm^{-1}, with the intensity of the band at 997 cm^{-1} too large relative to the intensity at 820 cm^{-1} to be the result only of crystalline MoO_3. After H_2O exposure, the spectrum (Figure 4.5c) contained band positions and intensities that were in agreement with those for heteropoly silicomolybdic acid (51–53). This type of surface silicomolybdic compound was previously inferred from other experimentation, including UV spectroscopy and acidimetric titration (54–56). The generation of silicomolybdic acid upon H_2O exposure of MoO_3/SiO_2 catalysts with loading from 2 to 8% MoO_3 indicated a change in the coordination of surface Mo from tetrahedral to octahedral and a change in the effective dispersion of Mo on the surface. Decomposition of silicomolybdic acid under calcination conditions previously had shown collapse of the Keggin unit (57,58). This surface structure was correlated to the mechanism of photocatalytic oxidation of propane (49,59).

Observation of a Raman band near 950 cm^{-1} for MoO_3/SiO_2 showed that a significant portion of the molybdena was contained in a highly dispersed phase when the MoO_3 loading was at or below 2 wt% (49). In another study of MoO_3/SiO_2, which involved Raman and FT-IR spectroscopy, X-ray diffraction, and photoluminescence, the amount of tetrahedrally coordinated molybdena was correlated with the rate of metathesis of propene (60). Here, the highly dispersed molybdena phase was described as an amorphous polymolybdate increasing in concentration up to a loading of approximately 1 atomic % MoO_3, after which its concentration decreased gradually. This highly dispersed phase also was said to contain an isolated species in the presence of crystalline MoO_3. Thomas et al. (61) also had described a highly dispersed molybdena species on SiO_2.

Seyedmonir and Howe (62) examined the redox chemistry of molybdena on silica catalysts. Their study employed a variety of spectroscopic techniques to detail Mo speciation and compared Raman vibrational frequencies of supported molybdena with standard compounds such as $(C_4H_9)_3SiOMoO_3^-$ (63). Changes observed in the characteristics of the Mo=O bands upon extended calcination and in situ analysis were found to be similar to changes previously observed (38). In addition, infrared analyses were performed on the same catalysts after calcination–hydration exposures of the catalysts. These infrared data were interpreted in terms of hydrolysis of Mo—O—Si bonds, the reverse reaction of that occurring when molybdic acid and support hydroxyl groups interact during catalyst synthesis. Incorporation of ^{18}O during exposure of Mo/SiO_2 catalysts to $H_2^{18}O$ occurred in the terminal Mo=O bonds.

MoO_3/TiO_2

Titania supports usually contain two distinct crystallographic structures, anatase and rutile. The intense Raman bands of these titania are located at 144, 199, 399, 520, and 643 cm^{-1}, and at 144, 148, and 611 cm^{-1}, respectively (17,18). As pointed out by Liu et al. (64), TiO_2 is well known for its photocatalytic activity in the partial oxidation of alkanes, alkenes, and alcohols (65–67). The formation of a surface molybdena was confirmed for MoO_3/TiO_2 by the use of laser Raman and UV–visible spectroscopy and corresponding photocatalytic activity measurements (64). For 50 m^2/g TiO_2, consisting of approximately 60% anatase and 40% rutile, the amount of MoO_3 calculated to achieve a monolayer of surface molybdena was approximately 4 wt% (4×10^{14} Mo atoms/cm^2). Even though the photo-efficiency of MoO_3/TiO_2 was five times less than that for pure TiO_2, the selectivity for suppressing secondary oxidation reactions was greater for MoO_3/TiO_2 than for pure TiO_2. The most intense Raman peak observed for the surface molybdena was near 940 cm^{-1} with 0.8% MoO_3; this peak shifted to higher frequencies with increasing MoO_3 concentration, having the position of a 967 cm^{-1} band with 4% MoO_3. As most recently discussed by Ramis, Busca, and Lorenzelli (68), the description of a surface monolayer on TiO_2 is at best a theoretical approximation, as CO adsorption experiments suggested that the TiO_2 support with 4% MoO_3 still did not behave as if it were uniformly covered by a monolayer of a surface phase. Instead, it is perhaps preferable to discuss a theoretical monolayer coverage, or to use some other terminology describing the specific loading of a surface species at which, if the species were uniformly covering the surface, a monolayer would be present.

Ng and Gulari (69) considered Mo═O stretching frequencies for MoO_3/TiO_2 catalysts at 955 cm^{-1} and lower to be the result of tetrahedral species, whereas Mo═O bands higher than 955 cm^{-1} were believed due to octahedral species. Differences in the Mo═O terminal stretching frequencies for this tetrahedral species relative to corresponding Mo═O stretching frequencies in aqueous solution were explained in terms of strong interactions between the molybdena surface layer and the TiO_2 support. Although these strong interactions were not thoroughly described, they were said to be responsible for the spreading out of crystalline MoO_3 on TiO_2 supports when mechanical mixtures of crystalline MoO_3 and TiO_2 were calcined (41). A similar spreading phenomenon had been discussed in the case of calcination of catalysts that contained crystalline MoO_3 on Al_2O_3 (26).

The hydrocracking activity for diphenylmethane over supported MoO_3 catalysts has been shown to be greater for molybdena on TiO_2 than for molybdena on SiO_2, ZrO_2, Al_2O_3, and MgO (70). The formation of a surface molybdate, as suggested by a broad band in the 940 to 970 cm^{-1} region (64,69), was further confirmed by the observation of relatively weak bands at 327 and 247 cm^{-1} (39). Together, these bands were assigned to a surface heptomolybdate. The reducibility of MoO_3/TiO_2 was greater than that for any of the other supports listed, perhaps because of the reducible nature of the support itself.

It is of interest to note that a shifting in the surface Mo═O terminal stretching frequencies as a function of dehydration of catalysts (38,49,64) was also observed to occur for MoO_3/TiO_2 catalysts (68). These results were obtained in infrared studies where the intense absorption of the support significantly obscured spectral acquisition in the region of the fundamental internal vibrational modes from the supported phase. It was difficult to observe vibrational modes of the supported phase by FT-IR—for example, in the 800 to 1000 cm^{-1} region intense absorption of the support generally eliminates quantitative infrared data acquisition; however, generation of overtones of the Mo═O stretching vibration allowed observation of such surface chemistry.

A quantitative study of Mo/TiO_2 catalysts (71) included a determination by Raman and XPS spectroscopy of the amount of highly dispersed, surface molybdena relative to the amount of crystalline MoO_3. The TiO_2, a 70% anatase–30% rutile mixture, had an initial surface area of 54 m^2/g. The concentration of molybdena at which crystalline MoO_3 started to form was 6 to 7 wt%. Only a surface interaction species was observed before this loading. Because the TiO_2 support contained well-defined bands, the Raman scattering cross section of the surface phase relative to the scattering cross section of crystalline MoO_3 could be determined. The average value of this ratio was 3.6 ± 0.4. This value should be compared to the same relative Raman scattering cross section for MoO_3/Al_2O_3, the value of which has been determined to be 17 (46). Reasons for this difference are not completely known, but imply a different absolute scattering cross section for a molybdena surface phase on Al_2O_3 and TiO_2. Importantly, these authors (71) showed that the relative Raman scattering cross section and the loading at which crystalline MoO_3 was detected were not dependent upon the hydration level of the catalyst.

Other Mo-Containing Catalysts

The use of Raman spectroscopy to study MoO_3 on other supports has just begun. It has been shown, for example, that the interaction of MoO_3 with CeO_2, MgO, ZrO_2, and SnO_2 produces a surface molybdena phase distinct from MoO_3 interacting with Al_2O_3 or SiO_2 supports (40,42). Of particular note was the generation of new bands for calcined MoO_3/MgO at 855, 907, 967, and 1094 cm^{-1}, with a complement of bands in the 250 to 400 cm^{-1} region, and for calcined MoO_3/SnO_2 at 799, 890, 976, and 1001 cm^{-1} (42). These spectra were obtained while using in situ cells or under careful control of the environment of the sample. In part, the observation of these bands and their shifting in frequency as a result of hydration–dehydration is reminiscent of changes observed upon in situ analysis of SiO_2 and Al_2O_3 supported MoO_3; however, an exact description of the type of surface species that produced such richness in band structure for MgO and ZnO supports has yet to be completely determined. In addition, the participation of these surface species, possibly different from those observed on Al_2O_3 or SiO_2, in catalytic reactions has yet to be elucidated.

The use of molybdenum-containing catalysts for olefin oxidation and ammoxidation has been studied by Callahan et al. (72). These catalysts (e.g., bismuth molybdate) are selective in the oxidation of olefins through the interaction of lattice oxygen and adsorbed olefin intermediates, and require one or more distinct active sites. These sites would initiate H abstraction and oxygen or NH insertion (73–75). The fact that this type of catalyst would be used in the oxidic or partially reduced form during catalysis implies an ease of Raman data acquisition for possibly catalytically important phases. Hence, Glaeser et al. (76) used Raman spectroscopy to investigate whether the lattice oxygen associated with Bi caused H abstraction, and whether the oxygen associated with molybdenum caused insertion into the allylic intermediate. Raman data were correlated with the results from reactor tests; the location of ^{18}O substitution into the Bi_2MoO_6 lattice after catalyst reduction also was examined by the Raman analysis of isotopic shifts in vibrational spectra of the molybdate. Characteristics of such ^{18}O-substituted Bi_2MoO_6 had been previously published (77,78), along with its Raman and infrared band assignments (79). These assignments were of fundamental importance because it was found that reduction of the catalysts by use of but-1-ene and subsequent ^{18}O isotopic substitution did not produce frequency shifts in the peaks at 853, 798, and 720 cm^{-1}; these bands were assigned to Mo—O stretching modes. Oppositely, catalyst reduction with propene, methanol, or ammonia and subsequent ^{18}O substitution caused isotopic shifts in the same Mo—O stretching modes. These data suggested that H abstraction was activated by lattice oxygen in the bridge position of Mo—O—Bi, whereas insertion into allylic intermediates involved only Mo—O bonds.

Kovats and Hill also investigated bismuth molybdate catalysts using Raman spectroscopy (80). Their results on $Bi_2Mo_3O_{12}$, $Bi_2Mo_2O_9$, Bi_2MoO_6, and $Fe_2Mo_3O_{12}$ focused on the origin of bands in the 1250 to 1600 cm^{-1} region as a function of catalyst reduction and oxidation. Although the positions of these bands, 1575 and 1375 cm^{-1}, were in close agreement with those expected from surface carbon (see below), experimental results indicated that their origin possibly was from low-energy charge transfer processes.

Summary of Supported Molybdenum

Raman spectroscopy has been particularly useful in investigating the surface speciation of supported Mo. In general, the types of surface species that have been observed include three-dimensional molybdenum trioxide (MoO_3), highly dispersed monomeric and polymeric molybdena anchored to the support surface, two- and three-dimensional molybdates, and molybdenum-support crystalline compounds. The type of Mo species that would be present depends on the degree to which Mo interacts with the support during impregnation and subsequent catalyst treatment. Although some Raman results on sulfided and/or reduced Mo species have been published, a considerable amount of Raman spectroscopic work is needed to provide an in-depth picture of these reduced or sulfided species.

The use of laser excitation wavelengths other than in the visible region may be helpful in obtaining good-quality Raman data on such black materials.

Supported MoO_3 with Co and Ni Promoters

CoO-MoO_3/Al_2O_3

Catalytic effects of Co and Ni promoters in MoO_3/Al_2O_3 catalysts have been studied extensively. For example, optimal Co/Mo ratios, the effect of preparation conditions, type of support, pH of impregnation, and other factors affecting their activity for hydrodesulfurization (HDS) of thiophene and of coal liquids have been studied (81–87). It has been suggested that the Mo is associated with desulfurization, while Co provides synergistic effects in both unsupported and supported catalysts (87–90). More recent studies of Co-Mo/Al_2O_3 catalysts show that Co interacts directly with Mo at edge sites in molybdenum islands (91–93).

The earliest laser Raman studies of these types of HDS catalysts were aimed at obtaining information about the surface structure that other experimental techniques could not present, and defining surface molecular speciation that was a factor in catalyst activity and selectivity. Hence, for example, the studies by Brown et al. (3,14) depicted new bands in the Raman spectra of CoO-MoO_3/Al_2O_3 that were not observed in MoO_3/Al_2O_3. The new bands, for catalysts having a constant MoO_3 loading of 7.5% or 15% and with CoO concentration varying from 0 to 9.5%, were located at 941 and 952 cm^{-1}. These bands were assigned to the presence of beta-$CoMoO_4$ (see Figure 4.6 and below). The intensity of the terminal Mo═O stretching vibration at 1006 cm^{-1} was substantially affected by the incorporation of Co, thereby suggesting an interaction of surface molybdena with CoO. The form of this interaction, especially in the highly dispersed surface phase, was not explicitly described; however, the detection of species such as beta-$CoMoO_4$, Co_3O_4, and MoO_3 as a function of Co/Mo ratios showed the utility of Raman spectroscopy for catalysis research.

The surface molecular speciation described by Medema et al. (15) was similar to that described by Brown et al. (3,14). In addition, Medema et al. (15) reported the presence of subsurface $CoAl_2O_4$ species. This type of Co^{2+} species had been previously discussed relative to XPS and TGA data and was suggested to facilitate the transformation of tetrahedral Mo to an octahedral polymolybdate. Formation of $Al_2(MoO_4)_3$ also was observed, and was defined to be located on top of any crystalline MoO_3 that had formed. The order and means of impregnation—whether catalysts were prepared by sequential impregnation (Co first, then Mo or Mo first, then Co), coimpregnation, or mechanical mixing of CoO with MoO_3—were also shown to affect surface speciation (31). In particular, the effect of Co was shown to form a more well-defined molybdena at lower MoO_3 concentrations than when Co was not present, as well as cause differences in

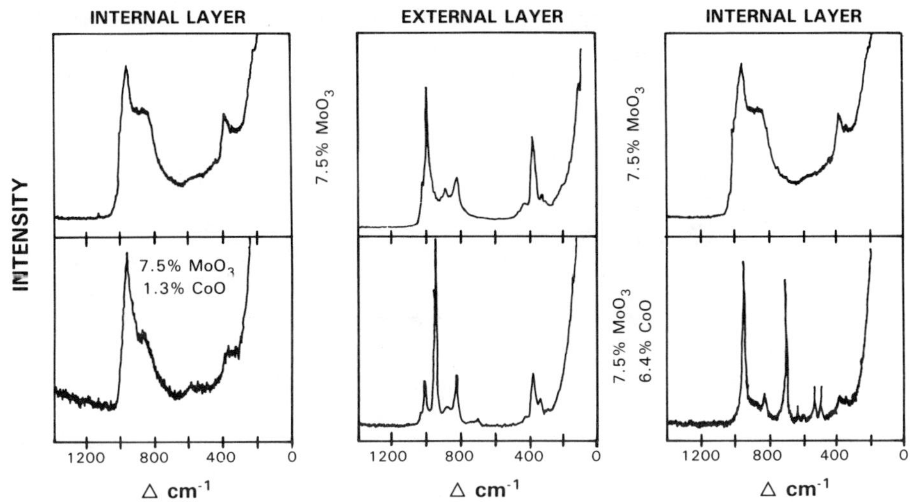

Figure 4.6. Raman spectra of the external and internal layers of mixed cobalt and molybdenum-on-alumina catalysts (internal layers monitored in the same manner as for Figure 4.1) (14).

the degree of aggregation of the surface molybdate. Crystalline MoO_3, usually detected at near-theoretical monolayer levels of MoO_3, was found only at concentrations of molybdena higher than monolayer capacity when Co was present.

Commercial Co-Mo/Al_2O_3 catalysts generally do not contain $CoMoO_4$ when the CoO loading is near 3 wt% CoO with 15 wt% MoO_3. With respect to such speciation, Chung and Massoth (40) showed that the sequence of incorporating Co and Mo onto Al_2O_3 affected surface speciation in these HDS catalysts. In the case of sequential impregnation, the calcination process in between impregnation steps was considered to firmly fix the Co or Mo to the Al_2O_3 surface, thereby decreasing the possibility of Co–Mo interactions that would enhance formation of $CoMoO_4$. In coimpregnation, such fixing of the surface species to the support was not possible; hence, $CoMoO_4$ could form. Chin and Hercules (95), by use of XPS and acoustic IR, did not find any evidence for the formation of $CoMoO_4$. However, when the Al_2O_3 support was thoughly dehydrated before the impregnation of Co and Mo, the Raman band width in the Mo═O terminal stretching region was determined to change dramatically over a small change of the CoO concentration (96), and then, with a loading of 4% CoO–15% MoO_3, distinct spectral splitting appeared that suggested the generation of $CoMoO_4$ (see Figure 4.7). At 5% CoO, X-ray diffraction data also showed $CoMoO_4$. Hence, the generation of surface species is dependent on the impregnation sequence, the pH of the impregnating solution, the surface properties of the support, and

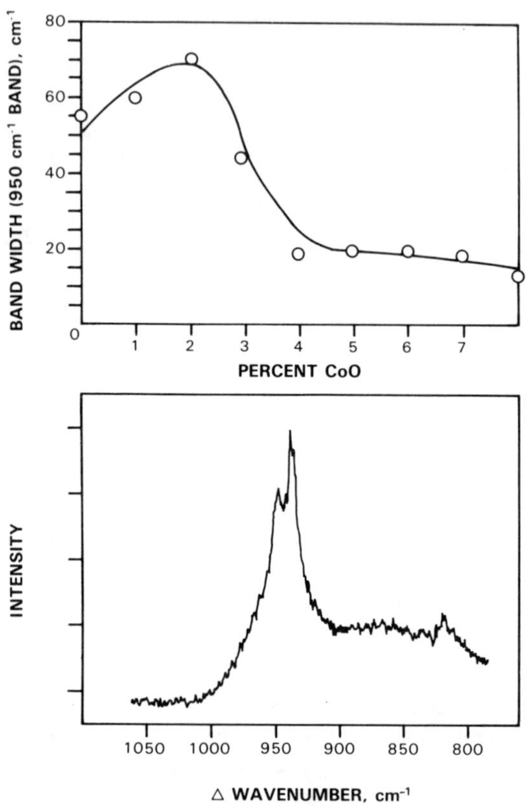

Figure 4.7. Raman results for CoO-MoO_3/Al_2O_3 as a function of CoO loading, in which the half width of the 950 cm^{-1} band changes and shows doublet band structure at loadings greater than 4% CoO; this doublet structure indicates formation of $CoMoO_4$ (96).

the relative concentration of promoter and active component when using a specific Al_2O_3 support.

The presence of molybdena with CoO on Al_2O_3 was suggested to decrease the potential for the formation of bulk Co_3O_4 (97); these researchers also discussed the formation of $CoMoO_4$ and $Al_2(MoO_4)_3$ with respect to the impregnation sequence and calcination temperatures of the catalysts. Impregnation of only Co and subsequent formation of bulk Co_3O_4, as monitored by Raman spectroscopy, was found to occur at Co concentrations as low as 1 wt% (15,16). When Mo was present, Co_3O_4 was observed only at higher CoO concentrations. The type of $CoMoO_4$ formed on these catalysts, if any, has been investigated to some degree (3,14,98). For example, at atmospheric conditions the two forms of $CoMoO_4$ are known as alpha, having tetrahedrally coordinated Mo, and beta,

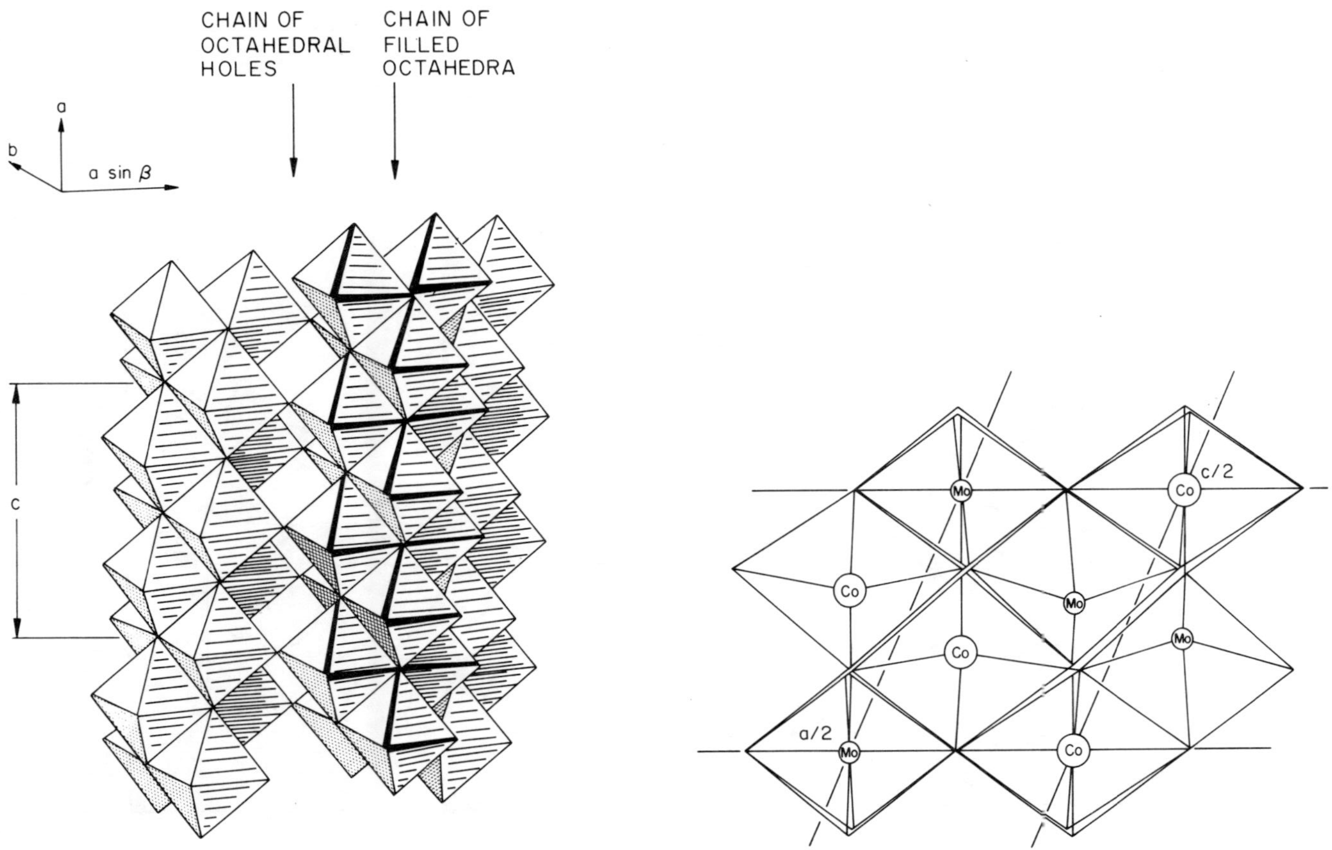

Figure 4.8. Diagram of the beta structure of $CoMoO_4$, in which the Co and Mo are octahedrally coordinated.

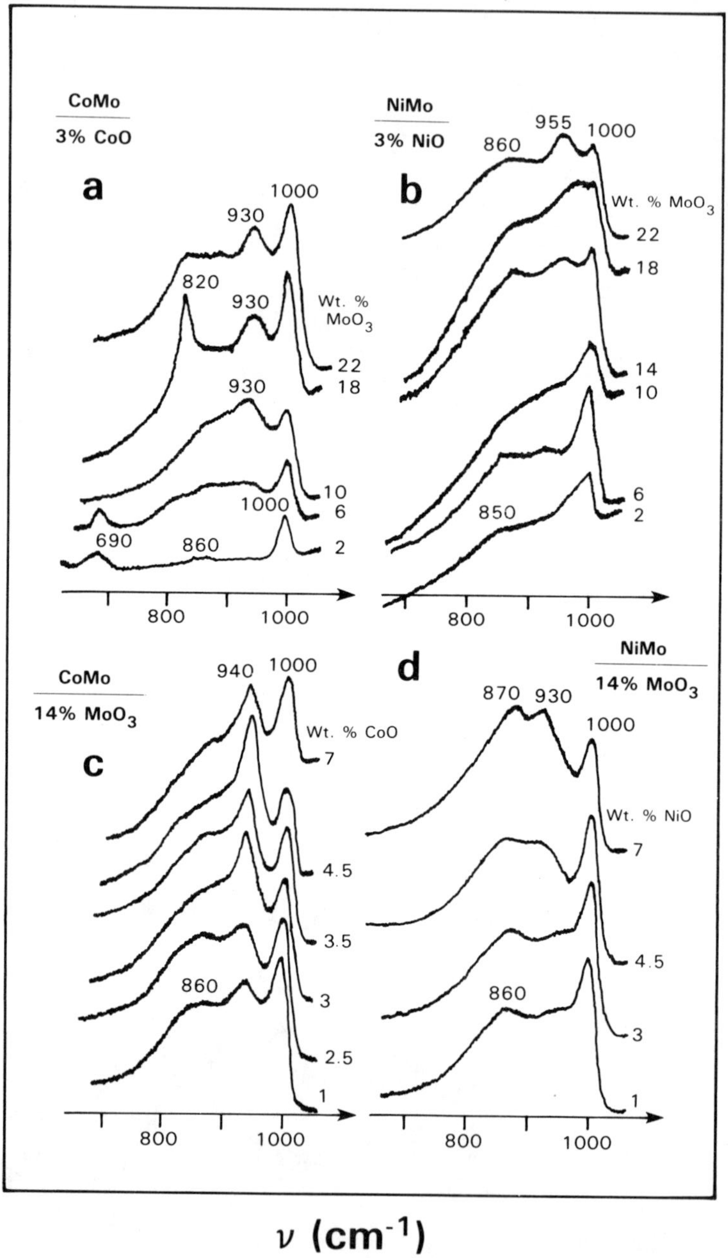

Figure 4.9. Raman spectra (high laser power ~ 300 mW) of Co-Mo and Ni-Mo samples calcined in air (500°C for 2 hours): (a) constant Co content (3% CoO), (b) constant Ni content (3% NiO), and (c) and (d) constant Mo content (14% MoO_3) (99).

having octahedrally coordinated Mo. In both the alpha and beta phases the Co is octahedrally coordinated. The beta structure is depicted in Figure 4.8. Characteristic Raman band positions of the alpha form are at 938, 820, and 700 cm^{-1}. The Mo═O stretching band in the beta form is split, containing two components located at 946 and 938 cm^{-1}, in addition to having relatively weak bands at 870 and 820 cm^{-1}. It is possible by the choice of appropriate conditions to transform alpha to beta and beta to alpha, and to follow the solid state reactions during the formation of these molybdates (98).

Spectral characteristics in the 700 to 1100 cm^{-1} region for calcined Co-Mo/ and Ni-Mo/Al_2O_3 catalysts are shown in Figure 4.9 (99). These spectra were obtained by using a controlled environmental cell, in which the exposure of the catalyst to gases could be controlled. The ca. 1000 cm^{-1} band was associated with terminal Mo═O stretching vibrations in dehydrated molybdena (33,99). An effect of the promoter was to increase the ease of formation of this band, even at low MoO_3 concentrations, as compared to its formation in MoO_3/Al_2O_3. Such increased ease of dehydrating the surface molybdena suggested that Co or Ni was located near the Mo, although an exact description of the structure was not given. With increasing MoO_3 concentrations the bands in the 930 to 940 cm^{-1} region increased in intensity relative to the 100 cm^{-1} band; these 930 to 940 cm^{-1} bands were attributed to hydrated surface species with the lower position of 930 cm^{-1} being associated with isolated molybdena species. It is also of interest to note that Ni promotion enhanced the intensity of the 870 cm^{-1} band relative to the intensity of the 930 to 940 cm^{-1} band.

Raman spectroscopy has not been as widely applied to the examination of sulfided Co-Mo/Al_2O_3 as to the oxidized form of these catalysts. Schrader and Cheng (100), in an extension of their earlier work on sulfided Mo/Al_2O_3 (31), suggested that the major effect of Co promotion was to increase the ease of reducing a surface oxysulfide phase. In addition, their results showed relatively weak MoS_2 peaks compared to what was observed for sulfided Mo/Al_2O_3; this indicates that the Co caused the breakup or prevented the creation of large MoS_2 crystallites. The presence of cobalt sulfides in these catalyst types has not been observed by Raman spectroscopy, probably because of the weak scattering cross section and black color of Co_9S_8. However, Payen et al. (98) have shown that the Raman bands for the MoS_2 phase in sulfided Co-Mo/Al_2O_3 are shifted in position relative to those in sulfided Mo/Al_2O_3. This shift was considered to be due to the intercalation of Co between the layers of MoS_2.

Raman spectral characteristics of thiophene and pyridine adsorbed onto Co-Mo/Al_2O_3 were also discussed by Payen et al. (98). No bands were observed that were similar to those for liquid thiophene, that is, at 832, 1031, 1035, 1358, and 1401 cm^{-1}. Instead, only bands in the 1300 to 1600 cm^{-1} region were observed subsequent to thiophene adsorption, these bands being due to carbonaceous deposits formed during laser irradiation (1–3). Other types of spectral analyses of thiophene adsorption (101,102) have been more informative than the Raman spectral acquisition in this particular case.

NiO-MoO$_3$/Al$_2$O$_3$

The addition of Ni to MoO_3/Al_2O_3 is advantageous in promoting the HDS and HDN of petroleum feedstocks (103,104). If highly dispersed surface species are formed in these catalysts, their Raman spectra may be expected to be identical to the Raman spectra obtained from analogous Co-promoted MoO_3 catalysts. However, comparison of Raman and XPS spectra from MoO_3/Al_2O_3 with those from NiO-MoO_3/Al_2O_3 and NiO/Al_2O_3 have provided important insight into the interaction of NiO with molybdena surface species (105–107). These spectra suggested subtle differences in the surface speciation of molybdena having Ni promotion as compared to that having Co promotion. Such differences were in excess of the effects of crystalline compound formation at specific Ni/Mo stoichiometries. In the case of the formation of crystalline compounds such as $NiMoO_4$, the spectra are expected to be very similar to the corresponding Co-promoted catalysts because $NiMoO_4$ is isomorphic with $CoMoO_4$ and also consists of alpha and beta forms.

Dufresne et al. (106) observed an intense Raman band near 885 cm^{-1} for Ni-Mo–containing catalysts; this band, the appearance of which was dependent on the state of the support surface, was assigned to a vibrational mode of a tetrahedral surface molybdena. Above 6 wt% MoO_3, the type of surface molybdate was described to be a polymolybdate containing its characteristic Raman band near 950 cm^{-1}. For NiO/Al_2O_3, Raman special characterization indicated the formation of a surface $NiAl_2O_4$ compound. This compound contains tetrahedrally and octahedrally coordinated Ni^{2+} ions, whereas Co in $CoAl_2O_4$ occupies only tetrahedral sites. Upon incorporation of molybdena and Ni onto Al_2O_3, the type of surface species formed was different from that formed when only molybdena was present. For example, at low Ni and MoO_3 concentrations, $NiAl_2O_4$ was formed; the interaction of Ni with Mo did not necessarily produce a $NiMoO_4$ surface compound, but if one were formed, the alpha form of $NiMoO_4$ was observed.

The existence of Ni^{2+} ions within the surface polymolybdate and in the Al_2O_3 for NiO-MoO_3/Al_2O_3 has also been postulated. The fact that no NiO or $Ni(OH)_2$ was observed in these Raman studies was recently addressed by Chan and Wachs (108). For NiO, which has the NaCl structure, the Ni is located in octahedral sites (109) and is relatively easily reduced to Ni metal as compared to the Ni in $NiAl_2O_4$ (110). The Raman band from crystalline NiO is located in the 400 to 550 cm^{-1} region, and may consist of two overlapping bands. For NiO/Al_2O_3, the Raman band associated with Ni=O stretching vibration was centered at 550 cm^{-1}. The intensity of the nickel–oxygen stretching vibration in crystalline and supported NiO is significantly weaker than the Mo=O stretching vibrations from surface molybdena. For example, on identical supports the Raman spectrum of 10 wt% NiO would have a lower signal-to-noise ratio than a corresponding 3 wt% loading of MoO_3. In the case of $NiAl_2O_4$, having Raman bands at 200, 375, and 600 cm^{-1}, band intensities are also relatively weak and would be difficult to observe in samples containing strong Raman scatterers such as molybdena. The main thrust of this research (108), was, however, to discuss the characterization of supported metal oxides that have high oxidation states, such as W^{6+}, Mo^{6+}, V^{5+}, Re^{7+}, and Cr^{6+}, and that contain intense M=O terminal stretching modes near 1000 cm^{-1}. These metal oxides form surface species that are dramatically influenced by the adsorption of H_2O. In the case of low-oxidation-state metal oxides, such as Ni^{2+}, Cu^{2+}, and Co^{2+}, the interaction of the ion with a support usually creates a subsurface species that is not sensitive to the adsorption of H_2O.

Highly active Ni-Mo/Al_2O_3 catalysts do not normally contain crystalline phases of $NiMoO_4$, $Al_2(MoO_4)_3$, or MoO_3. Instead, as described by Kasztelan et al. (111) and Bachelier et al. (112,113), high activity for the HDS of thiophene over Ni-Mo/Al_2O_3 was the result of highly dispersed surface molybdenum–sulfur islands, edge-decorated by Ni atoms. These so-called NiMoS sites are similar in physical definition to CoMoS sites in sulfided Co-Mo/Al_2O_3 (91–93). The presence of crystalline components in the oxidic form of these catalysts will decrease their HDS activity. Similarly, it may be expected that the generation of highly active Mo–sulfide islands in these HDS catalysts may depend on the form of the highly dispersed surface molybdate. Such a possibility was investigated by use of Raman spectroscopy and ISS (114). In this study, the influence of hydration–dehydration cycles on the surface molybdate was discussed in relation to the influence of Ni promotion. The effect of Ni on the intensities of the terminal Mo=O stretching band near 1000 cm^{-1} as compared to intensities of Mo-O-Mo or Mo-O-Al stretching bands in the 700 to 900 cm^{-1} region is shown in Figure 4.10. The trend in these data, having a discrete inflection point that is dependent on the overall molybdena concentration, suggested a direct interaction of Ni with terminal Mo=O bonds. In addition, the Ni/Mo atom ratio at which the inflection point occurred was in excellent agreement with the Ni/Mo atom ratio at which a maximal promotional effect of Ni was observed for the HDS of thiophene (112,113).

Whether the characterization of the oxidic form of a catalyst can be used to provide insight into properties of an active, sulfided phase of the catalyst has been a topic of general interest in catalysis (115). Comparison of recent spectroscopic results for oxidic forms of Ni- and Co-molybdena catalysts and corresponding reactor tests (116,117) indicate a significant and applied interpretation of the spectroscopic data from their oxidic form to the active-sulfided form. Hence, the data in Figure 4.10 may be of value for defining the efficacy of the sulfided form of Ni-Mo/Al_2O_3. The difference in the value of the intensity ratio

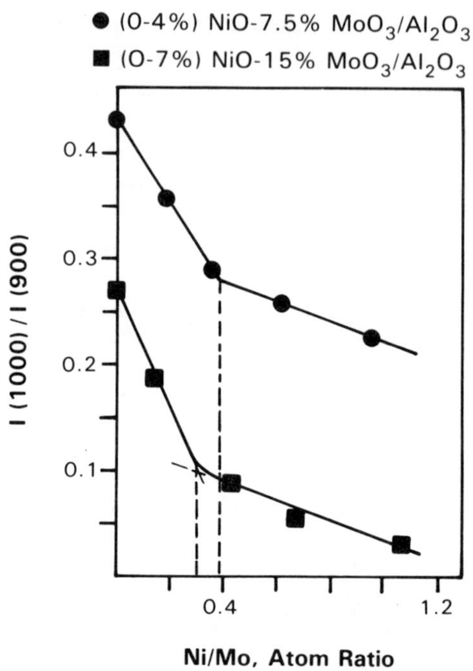

Figure 4.10. The integrated intensity of the 1000 cm^{-1} band, $I(1000)$, ratioed to the integrated intensity of the band in the region 850 to 950 cm^{-1}, $I(900)$, as a function of NiO concentration for 500°C O_2-calcined catalysts (114).

in Figure 4.10 for the 7.5% and 15% MoO_3 samples also may be indicative of differences in the condensed nature of the supported phase. For example, TGA analyses of $MoO_3Al_2O_3$ have shown that the hydration capacity of MoO_3/Al_2O_3 was larger than for gamma-Al_2O_3, with an approximate increasing extent of hydration proportional to the Mo content (~ 2 mol H_2O/MoO_3) (116). Such a dependence was in agreement with the Raman spectral results of Payen et al. (118), who discussed the assignment of a broad 850 cm^{-1} band in these catalysts as a function of hydration–dehydration. In particular, their assignment of the 850 cm^{-1} band to Mo-O-Mo bridging groups within the surface molybdate confirms the interpretation of the data shown in Figure 4.10 (114). The position and intensity of the broad band in the 800 to 900 cm^{-1} region was not sensitive to hydration–dehydration cycles (114,115). Its intensity was, however, sensitive to the concentration of molybdena, indicating that at low concentrations the molybdena surface species was isolated in nature. At higher concentrations (~ 7% MoO_3), increasing intensity of this band suggested the development of a more condensed structure for the surface species.

Kantschewa et al. (119) used Raman, diffuse reflectance, XPS, and infrared spectroscopy in conjunction with reactor conversion tests to elucidate the effect of K addition to Ni-Mo/Al_2O_3 catalysts. The destruction of octahedral surface molybdena and the concomitant creation of tetrahedral surface molybdena was dependent on the concentration of the added K. This type of Mo-based catalyst previously had been tested for activity in the water gas shift reaction (120,121). The presence of K created two separate molybdena phases: K_2MoO_4 and $K_6Mo_7O_{24}{\cdot}4H_2O$. Calcination of the catalysts at 250°C caused distortion of the adsorbed tetrahedral MoO_4^{2-} anion, thereby increasing the M=O stretching vibrational frequency from 894 to 916 cm^{-1}. Although HDS activity was significantly decreased by the addition of K, these K-promoted Ni-Mo/Al_2O_3 had water gas shift activity that was greater than that without K. The Raman results thereby provided complementary spectroscopic information on surface species of importance in catalytic activity.

Summary of Promoted Mo/Al_2O_3 Catalysts

Raman spectroscopic analysis of these promoted catalysts has provided extensive knowledge about the association of promoters with the molybdenum and support phase. Most of this information has been obtained for the oxidic form of these catalysts and not the active sulfided form. Although there may be a direct correlation of surface and bulk molecular speciation in the oxidic form with the sulfide form, it would be advantageous to show such a correlation by a comprehensive spectroscopic and reaction testing program. It is possible to obtain good-quality Raman spectra from black, sulfided catalysts. However, alternate excitation sources may be needed to increase the quality of the Raman data. In particular, a substantial decrease in fluorescence would be expected with the use of infrared excitation as compared to visible excitation.

Supported WO_3

WO_3/Al_2O_3

The type of surface species generated when tungsten is impregnated onto Al_2O_3 has been extensively studied by the use of Raman spectroscopy. The interpretation of these spectral data has relied on the transfer to the tungsta system of infrared and Raman spectral data on otherwise well-characterized compounds. For example, the tetrahedral ion, WO_4^{2-}, in aqueous solution contains four fundamental vibrational modes denoted by and having symmetry of $\nu_1 (A_1) + \nu_2 (E) + \nu_3 (F_2) + \nu_4 (F_2)$; all four of these modes should be Raman-active (122). However, only three Raman bands have been observed, having frequencies of 931, 325, and 833 cm^{-1}, respectively. The positions of ν_2 and ν_4 are considered to be located within ± 10 cm^{-1} of each other at 325 cm^{-1} (123).

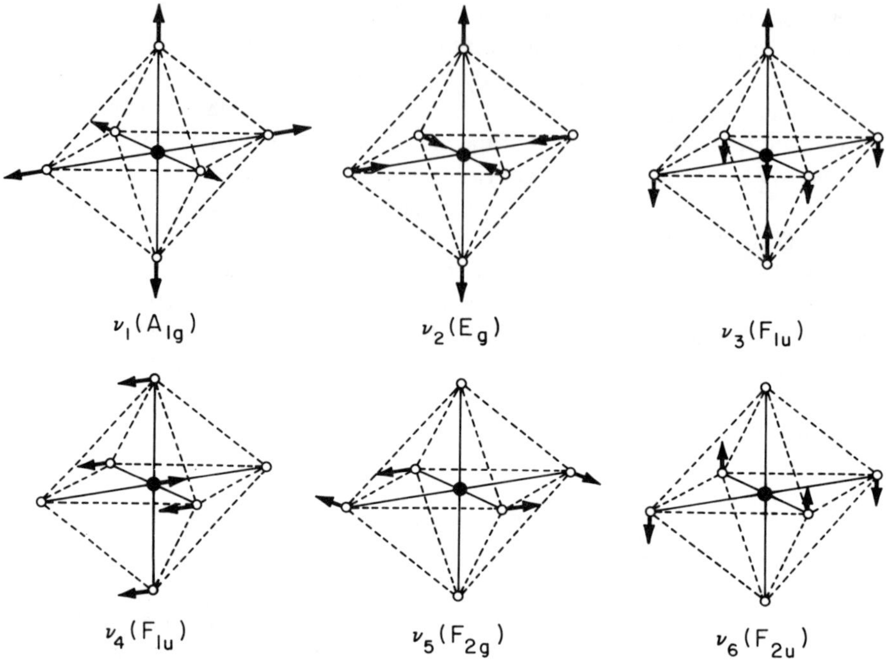

Figure 4.11. Vibrational modes of octahedral WO_6 species.

Octahedrally coordinated tungstate anion, WO_6, possesses only three Raman-active vibrational modes, denoted by and having symmetry of ν_1 (A_{1g}) + ν_2 (E_g) + ν_5 (F_{2g}). In either octahedral or tetrahedral anions any distortion of the W-O bond angles causes a lowering of the symmetry of the species and the possible generation of additional Raman activity.

The above description of octahedral symmetry contains singly degenerate, E, and doubly degenerate F, vibrational species. Vibrational motions associated with these modes are shown in Figure 4.11. Lowering of molecular symmetry can be assumed to remove this degeneracy with the concomitant creation of band splittings and an increase in the number of modes for the WO_6 species to $3N - 6 = 15$, where $N = 7$. An example of an octahedral species is crystalline WO_3, which contains Raman vibrational bands at 807, 712, and 263 cm^{-1} (124); besides these three bands, there are numerous bands in the low frequency region of the Raman spectrum of crystalline WO_3 that are assignable to lattice vibrations of the crystal. Another compound containing octahedrally coordinated tungsten is Li_6WO_6 (125); here, the three Raman modes are located at 740 cm^{-1} (A_{1g}), 430 cm^{-1} (E_g), and 360 cm^{-1} (F_{2g}). An example of tetrahedral tungsten would be Na_2WO_4 (123,126), which has Raman vibrational modes located at 928 cm^{-1}

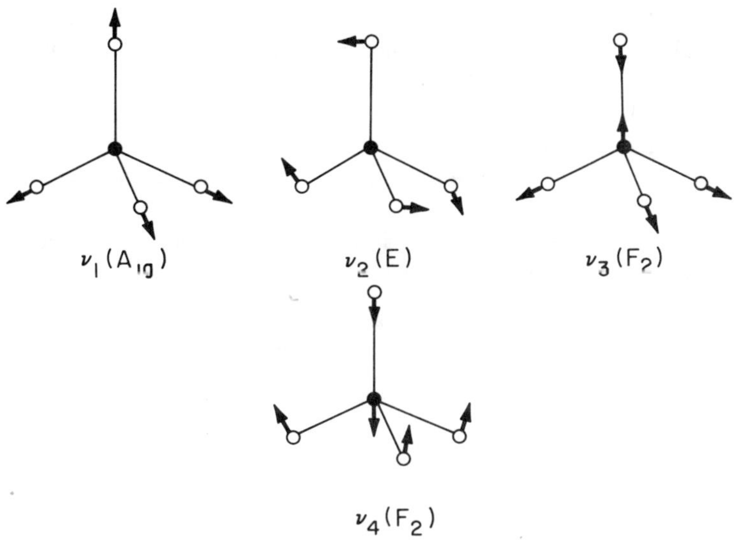

Figure 4.12. Vibrational modes of tetrahedral WO_4 species.

(A_{2g}), 312 cm^{-1} (E), 813 cm^{-1} (F_2), and 373 cm^{-1} (F_2). Vibrational motions associated with these modes are shown in Figure 4.12. Other examples of compounds that have tetrahedrally coordinated tungsten are $PbWO_4$, $CaWO_4$, and so on; these compounds contain what has been termed the scheelite structure, and in some cases contain distorted tetrahedra (127–129).

The chemistries of Mo and W are closely related, so it can be expected that the surface molecular speciation of supported molybdena and tungsta would also be closely related. In fact, this is the case. As early as 1973, Biloen and Pott investigated the reduction of supported tunsgsten with the use of XPS techniques and found that the Al_2O_3 support stabilized tungsta against reduction as compared to the reduction of bulk compounds such as WO_3 (130). Thomas and co-workers (131–133) studied WO_3/Al_2O_3 by use of Raman spectroscopy for WO_3 concentrations of 2 to 29 wt%. Their data showed that the frequency of the most intense band increased by 35 cm^{-1} as the concentration of WO_3 increased. At the higher WO_3 loadings, the position of the most intense band was 973 cm^{-1}, which was assigned to the W-O stretching vibration of octahedrally coordinated W. This assignment was based on the single band at 973 cm^{-1}, in correspondence with Raman research that had been performed on MoO_3/Al_2O_3 and by comparison of their data with Raman activity from well-characterized solid and aqueous compounds.

The XPS data (130) for reduced WO_3/Al_2O_3 suggested the presence of $Al_2(WO_4))_3$. In contrast, Iannibello, Villa and Marengo (134) did not observe such compound formation, but, in agreement with Thomas et al. (131–133), they found a well-dispersed oxo-tungsten species that produced a broad and intense band in the 965 to 990 cm^{-1} region. This band also shifted to higher frequencies with increasing WO_3 concentration. The coordination of this species was assumed to be tetrahedral. Distortion of the T_d symmetry of WO_4^{2-} was considered to cause an increase in its double bond character, thereby increasing its vibrational frequencies.

The structure of the tungstate ion in aqueous solutions is pH-dependent, according to the following equilibrium (135,136):

$$12WO_4^{2-} + 12H^+ \leftrightarrows W_{12}O_{42}^{12-} + 6H_2O$$

$$pH > 7 \qquad\qquad\qquad pH < 7$$

Raman bands of tetrahedral WO_4^{2-} are the result of the W=O symmetric (ν_1, 931 cm^{-1}) and antisymmetric (ν_3, 830 cm^{-1}) stretching modes and W=O bending modes (ν_2, ν_4: 324 cm^{-1}). For $W_{12}O_{42}^{12-}$, the major Raman bands are located at 977, 963, and 166 cm^{-1} (137), with the lowest band position attributed to the W-O-W deformation mode. Upon bond angle distortion and/or interaction with an Al_2O_3 surface, the vibrational positions of these tetrahedral and octahedral species can increase. Hence, Salvati et al. (138) interpreted the Raman spectra of WO_3/Al_2O_3, which contained bands at 973 and 333 cm^{-1}, to be the result of a tetrahedrally coordinated tungsta species; such a species would have experienced a 4% increase in its W=O stretching and bending modes as compared to the corresponding modes in aqueous WO_4^{2-}.

The XPS and ISS data on WO_3/Al_2O_3 (190 m^2/g Al_2O_3) indicated that a WO_3 concentration of 24 wt% formed a monolayer of tungsta (138). Raman bands at 807, 710, and 272 cm^{-1} for catalysts containing 15 to 24 WO_3 suggested the formation of octahedral, crystalline WO_3. However, the species associated with these bands was significantly less reducible than crystalline WO_3 (139), thereby requiring that these Raman bands be assigned to those of an octahedral WO_3-like interaction species. Above a concentration of 24% WO_3 (the approximate monolayer level), Raman data showed increasing intensities of bands associated with crystalline WO_3. Coincident with these increasing Raman intensities was an increasing reducibility of the tungsta species to W^0. Hence, the tungsten speciation was defined as tetrahedral up to a loading of 65% of the theoretical monolayer, and octahedral between this loading and the theoretical monolayer, and then included three-dimensional WO_3. No $Al_2(WO_4)_3$ was detected; in general, $Al_2(WO_4)_3$ has been detected in only a few studies (131).

No shifting in the terminal W=O stretching mode as a function of tungsta concentration was observed in the study of Salvati et al. (138). Other studies showed that this stretching vibrational mode was sensitive to hydration–dehy-

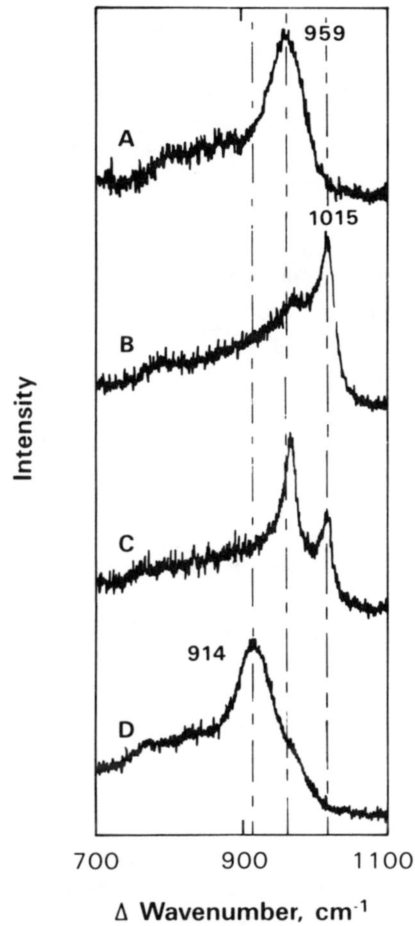

Figure 4.13. Raman spectra of 15% WO_3/Al_2O_3 for (a) the catalyst after exposure to atmospheric conditions for four months, (b) ^{18}O exchange on the catalyst at 760 Torr of $^{18}O_2$ at 500°C for 20 hours, (c) ^{18}O exchange at 760 Torr of $^{18}O_2$ at 500°C for 70 hours, and (d) the ^{18}O-exchanged catalyst after exposure to 10 Torr of $H_2{}^{18}O$ at 25°C for 2 hours (142). (Reprinted by permission of John Wiley & Sons, Ltd.)

dration of the catalysts (140–143). For example, in Figure 4.13 spectral changes are depicted from some WO_3/Al_2O_3 catalysts upon calcination in $^{18}O_2$ and subsequent exposure to $H_2{}^{18}O$. The first step in calcination was described as a dehydration of the surface species, followed by exchange of ^{16}O by ^{18}O. Such exchange should decrease the terminal W═O stretching frequency by the factor $(16/18)^{1/2}$. This indeed was the case, as the 1015 cm^{-1} band for the dehydrated catalyst shifted to 960 cm^{-1} upon ^{18}O exchange, while the W═O bending mode at 350 cm^{-1} for the dehydrated catalyst shifted to 329 cm^{-1}. Subsequent $H_2{}^{18}O$ exposure to the ^{18}O-exchanged catalyst shifted this 960 cm^{-1} band to 914 cm^{-1}, whereas $H_2{}^{16}O$ exposure only decreased the frequency of the terminal W═O stretching mode to 950 cm^{-1}. The ISS data and the insensitivity of the position of the W-O-Al stretching band in the 850 cm^{-1} region to ^{18}O exchange showed that the ^{18}O exchange was mostly limited to terminal W═O bonds. Such an effect implied that Raman spectra from these catalysts were indeed the result of monitoring surface species. Some other ISS experimentation on WO_3/Al_2O_3 has further confirmed the sensitivity of Raman spectroscopy used to monitor tungsta surface species (144).

Generally, the frequency of the terminal metal–oxygen stretching mode is higher in WO_3/Al_2O_3 catalysts than in MoO_3/Al_2O_3 catalysts (138,142–145). The Raman spectral characteristics of H_2-treated WO_3/Al_2O_3 was briefly discussed by Payen et al. (44). Here, the intensity of the ca. 1000 cm^{-1} W═O terminal mode decreased upon H_2 treatment at 450°C, with a relative increase in the intensity of a band in the 800 to 950 cm^{-1} region. Subsequent O_2 exposure to the H_2-treated catalysts restored the Raman spectra to the original calcined character.

In an interesting publication on WO_3/Al_2O_3 catalysts, Horsley et al. (146) showed the benefits of using Raman spectroscopy with X-ray absorption near-edge spectra (XANES). The 2 and 10% WO_3 catalysts, supported on 180 m^2/g gamma-Al_2O_3, were shown to contain different amounts of tetrahedral species in monomeric and dimeric forms, and in octahedral surface species. In addition, the effect of hydrating the calcined catalysts was defined as increasing coordination of the surface species from tetrahedral to octahedral. Stencel et al. (142), by use of only Raman spectroscopy, could not discern whether frequency shifts associated with hydration–dehydration were the result of changes in coordination or of changes in the bond order of the terminal W═O bonds during H-bonding of H_2O.

As pointed out by Horsley et al. (146), the discrimination between tetrahedral and octahedral tungstate by use of a characteristic position of the highest frequency band is difficult, even though, in general, this band position would reflect the highest bond order in the sample. For example, the frequency of this band is listed in Table 4.4 for a number of compounds. Considerable overlap in frequencies is observed, although only octahedral W contain bands below 910 cm^{-1}, and only tetrahedral W contain bands above 980 cm^{-1}. For compounds containing octahedral and tetrahedral species, for example, $Na_2W_2O_7$ (146–148), the Raman spectra should contain bands that reflect the stoichiometry of the species of different coordination. Intensities of WO_4- and WO_6-associated bands from crystalline compounds suggest a relatively weaker intensity contribution from WO_6 components as compared to WO_4. However, this trend may not hold

Table 4.4. Highest frequency (cm^{-1}) of Raman bands for tetrahedral and octahedral tungsten oxide compounds.

OCTAHEDRAL COMPOUNDS	CM^{-1}	TETRAHEDRAL COMPOUNDS	CM^{-1}
Li_6WO_6	740	$CaWO_4$	913
WO_3	805	Cs_2WO_4	920
$FeWO_4$	856	Na_2WO_4	928
$CoWO_4$	886	$BaWO_4$	940
$MnWO_4$	886	$Fe_2\ (WO_4)_3$	1026
$NiWO_4$	893	$Al_2\ (WO_4)_3$	1060
$CuWO_4$	908	$PbWO_4$	905
H_2WO_4	951	$(WO_4)^{2-}$	931
$(NH_4)_6H_2\ (W_3O_{10})_4$	980		

for highly dispersed species. For example, an irreducible WO_3-like species having octahedral coordination, discussed by Salvati et al. (138), was more easily observed by Raman spectrascopy than was the corresponding tetrahedral species. The interaction of MoO_3 and WO_3 with surfaces also produces surface species of tetrahedral and octahedral coordination that have Raman scattering cross sections up to two orders of magnitude smaller than the corresponding octahedra in crystalline MoO_3 and WO_3 (46,140).

NiO-WO$_3$/Al$_2$O$_3$

Few Raman studies of Ni-W/Al_2O_3 catalysts have been reported. It was shown that the HDS of thiophene was promoted by the addition of Ni to the catalyst, such that the maximum promoting effect was at an atom ratio of Ni/(Ni + W) of $\sim$ 0.3 (148,149). This value is close to the value found for Ni-Mo catalysts (111–113). The XPS investigations of NiO-WO_3/Al_2O_3 (138,150,151) indicated that the presence of Ni^{2+} incorporated into the surface tungsta enhanced the reducibility of the surface tungsta.

The work of Iannibello et al. (134) on NiO-WO_3/Al_2O_3 indicated that Ni did not appreciably alter the Raman spectra from those observed for WO_3/Al_2O_3; shifts in the 970 cm^{1-} W═O stretching mode as a function of tungsta content were also similar to those observed for WO_3/Al_2O_3. However, a more recent Raman study of NiO-WO_3/Al_2O_3 showed that air-exposed and hydrated catalysts contained W═O stretching frequencies that were dependent on the NiO concentration (152). This shift in frequency implied an increased degree of hydration of the W═O units as a function of increasing concentration of NiO. In addition, dehydration of the catalysts, having concentrations of zero to 11% NiO/20% WO_3, produced a characteristic W═O stretching mode at 1010 cm^{-1}. Residual band intensity in the 800 to 950 cm^{-1} region was an increasing function of NiO

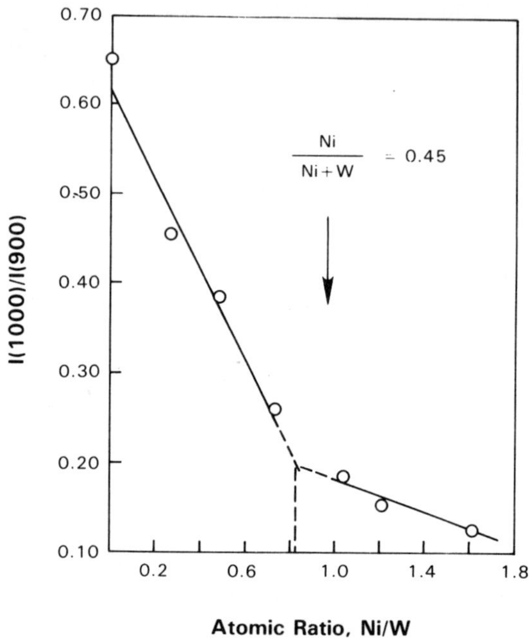

Figure 4.14. The integrated intensity of the 1000 cm^{-1} band, $I(1000)$, ratioed to the integrated intensity of the band in the 850 to 950 cm^{-1} region, $I(900)$, as a function of the Ni/W atomic ratio for NiO-WO_3/Al_2O_3 catalysts.

concentration. A plot of intensity ratios, as was done for NiO-MoO_3/Al_2O_3 in Figure 4.10, is displayed in Figure 4.14. It contains an inflection point at a Ni/(Ni + W) atomic ratio of 0.45. This atomic ratio was also the atomic ratio at which a maximum in the hydrogenation activity for coal liquefaction occurred (see Figure 4.15). Hence, the interaction of Ni with the surface tungsta species may be similar to that observed in Ni-Mo catalysts. Definitely, the Raman data in Figure 4.14 show a direct interaction of Ni with terminal W═O bonds.

WO_3/SiO_2

Investigations of WO_3/SiO_2 have indicated the presence of at least two distinct tungsta species (153,154), crystalline WO_3 and a highly dispersed surface tungsta. The surface species was identified as a polymeric, distorted octahedral complex.

Summary of Raman Results on Tungsta Catalysts

Raman spectroscopic data on supported tungsten catalysts have been unique in providing surface- and bulk-phase molecular information about the interaction of tungsta with supports and of promoters with the tungsta. Concentrations of

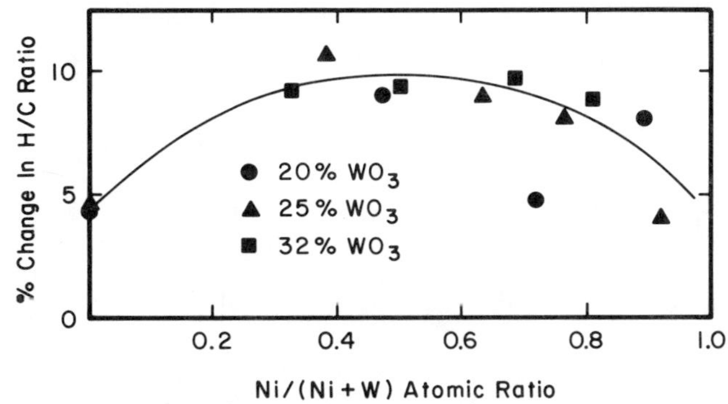

Figure 4.15. Hydrogenation activity of sulfided Ni-W/Al_2O_3 catalysts as a function of the Ni/(Ni + W) molar ratio. The maximum in activity coincides with the break in Raman intensity ratios plotted in Figure 4.14.

these phases have been shown to be quantitative and directly comparable to data from other spectroscopic techniques. Whereas most surface-sensitive analyses have not been able to delineate Ni-W, Ni-Mo, or Co-Mo interactions that relate to catalytic activity, the analysis of these catalysts by Raman spectroscopy has provided insight into the possibility that improved experimental techniques would allow active phase identification. Such studies are under way, having potentially exciting outcomes.

Supported Rhenia

Re_2O_7/Al_2O_3

Rhenia-containing catalysts are quite active for the metathesis of alkenes (155), whereas the addition of Re to Pt/Al_2O_3 catalysts has been found to produce highly active bimetallic reforming catalysts (156–158). The coverage of Al_2O_3 by use of high rhenium concentrations (~26 wt%) has been shown to eliminate characteristic alumina hydroxyl groups with the concomitant formation of rhenium-bonded OH (159). A laser Raman investigation of Re_2O_7/Al_2O_3 (160) concluded that the rhenia consisted of only tetrahedral ReO_4^- ions in weak interaction with the support. This assignment was the result of excellent agreement of the Raman spectra of (6–18%) Re_2O_7/Al_2O_3 with the corresponding spectra of NH_4ReO_4 and aqueous ReO_4^- (115,161–163). For NH_4ReO_4, the major Raman bands are located at 965 cm^{-1} (ν_s-ReO), 911 and 890 cm^{-1} (ν_{as}-ReO), and 339, and 332 cm^{-1} (O-Re-O deformation), whereas in aqueous solution the corresponding bands are at 970, 916, and 332 cm^{-1}. Frequencies of the supported species were at 968, 916, and 328 cm^{-1}, and were observed at these positions independent of rhenia concentration.

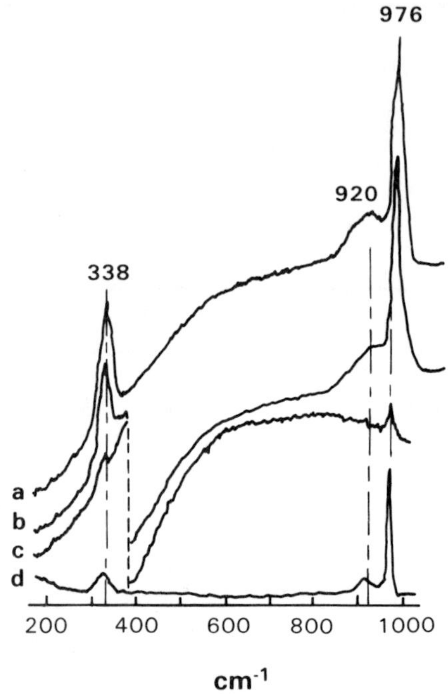

Figure 4.16. Raman spectra taken in air of calcined (500°C): (a) ReA-0.3, (b) ReA-1.2, and (c) ReA-2.9, made at pH = 9.0, 4.4, and 4.2, respectively. Spectrum (d) was from a 0.1 *M* solution of ammonium perrhenate (NH_4ReO_4). The ReA-2.9 sample was made from a 0.5 *M* solution. Its higher loading can be accounted for solely on the basis of the excess solute held in the pore system (167).

Yao and Shelef (164) suggested that the type of rhenia species for low Re_2O_7 loadings (~1.2–5.5%) included a two-dimensional surface phase and three-dimensional crystallites. Three-dimensional Re_2O_3 also was observed by Nakamura, Abe and Echigoya (165), but only after a loading of 17 wt% rhenia. A loading of around 20 wt% ($>5 \times 10^{20}$ Re/g) was calculated to be the level at which a theoretical monolayer of rhenia would form on 200 m^2/g Al_2O_3 (166). Hence, Wang and Hall (167) used their equilibrium adsorption method to impregnate rhenia onto Al_2O_3, ensuring that during impregnation the isoelectric point of the support was not exceeded. The Raman data showed the presence of only monomeric, tetrahedral ReO_4^- at coverages of up to 60% of the monolayer capacity. Corresponding visible adsorption spectroscopic data on these catalysts showed that considerable distortion of the symmetry of ReO_4^- existed. Such distortion, from T_d to C_{2v} or from T_d to C_{3v}, did not, however, cause the appearance of T_d symmetry forbidden Raman bands.

Wang and Hall (167) also examined what effect hydration–dehydration of Re_2O_7/Al_2O_3 had on their Raman spectra. These dehydration effects, depicted in Figure 4.16, include an increase in 976 cm^{-1} Re═O stretching mode to 1015 cm^{-1}, splitting of a 338 cm^{-1} band in the hydrated samples to 313 and 343 cm^{-1}, and the observation of new bands at 800 and 895 cm^{-1}. The increase of the Re═O frequency upon dehydration was said to be associated with the removal of H_2O, with a corresponding increase of the terminal Re═O bond order (39,114,140–143). The 800 cm^{-1} band in the Raman spectra of these alumina-supported rhenia catalysts was attributed to Al-O-Re vibrations.

Supported Vanadia

V_2O_5/TiO_2

Bulk vanadia is active in the selective oxidation of hydrocarbons; when supported on TiO_2, it has superior activity as compared to bulk V_2O_5 (168–170), with the preferred catalyst for the partial oxidation of *o*-xylene to phthalic anhydride being titania-supported vanadia. Raman characterization of freshly prepared V_2O_5/TiO_2 has suggested that two types of vanadia are present: a surface vanadia coordinated to the support and crystalline V_2O_5 (171). Extensive V-TiO_2 interactions occur upon high-temperature treatment of these catalysts. For example, it has been shown that as surface vanadia becomes reduced, it initiates the transformation of anatase to rutile, and is also incorporated into the TiO_2, forming $V_xTi_{1-x}O_2$ (171–174). The Raman spectrum of anatase is well known, having major peaks at 643, 520, 399, 199, and 144 cm^{-1}; the major peak of crystalline V_2O_5 is at 997 cm^{-1}, and is assignable to a V═O stretching mode (124,143).

Saleh et al. (175) and Wachs et al. (176,177) examined in close detail the speciation of V_2O_5/TiO_2 catalysts that had been prepared by oxalic acid impregnation. Laser Raman results as a function of catalyst calcination, depicted in Figure 4.17, show the creation of the 997 cm^{-1} band at temperatures near 200°C (concomitant with destruction of the precursor), and the disappearance of V_2O_5 after temperatures of 575°C were reached. Conversion of *o*-xylene also was shown to depend on catalyst calcination temperature. Within the temperature range 350 to 575°C, the surface area of the TiO_2 decreased with a corresponding increase in the Raman intensity of crystalline V_2O_5; within the temperature range 575 to 750°C, the solid state reaction of V_2O_5 with the TiO_2 support occurred (172–174). Hence, the data in Figure 4.17 signify the destruction of crystalline V_2O_5 and the formation of $V_xTi_{1-x}O_2$.

The distinction by Raman spectroscopy of crystalline V_2O_5 and V_2O_5 surface species was discussed on the basis of bands in the 900 to 1000 cm^{-1} region

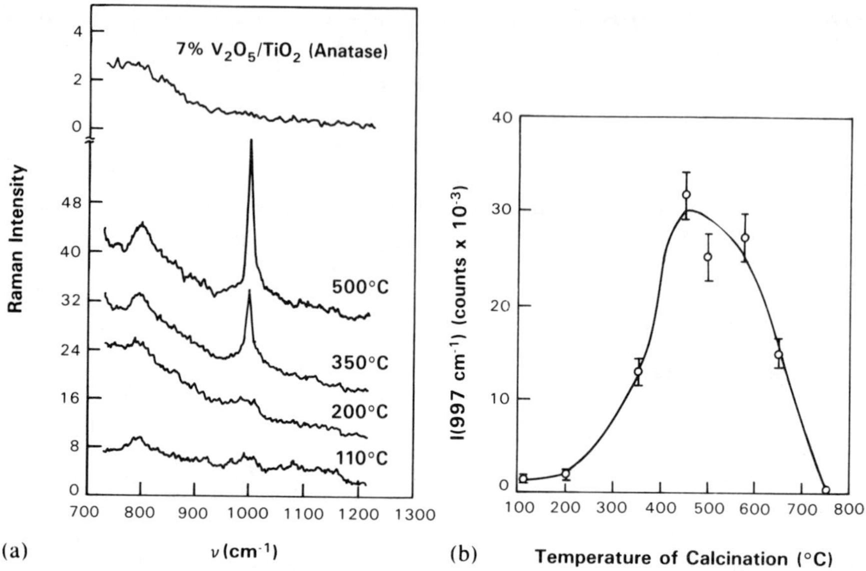

Figure 4.17. (a) Laser Raman spectra of 7% V_2O_7/TiO_2 (anatase) as a function of calcination temperature. (b) Intensity of 997 cm^{-1} band of bulk V_2O_5 for 7% V_2O_5/TiO_2 (anatase) as a function of calcination temperature (175).

(177). For example, at V_2O_5 concentrations less than that for theoretical monolayer coverage on 8.5 m^2/g anatase, assuming 0.105 nm^2 per $VO_{2.5}$ unit, the Raman spectrum contained a broad band centered near 920 to 970 cm^{-1}. When the approximate monolayer loading of 1.3 wt% V_2O_5 was exceeded, the 997 cm^{-1} band of crystalline V_2O_5 was observed. These Raman data were also examined relative to the activity of the catalysts for the oxidation of *o*-xylene (177), and an excellent correlation was found between conversion activity and surface speciation.

Hydrated surface vanadia on TiO_2 was shown to contain two relatively broad bands at 795 and 930 cm^{-1}, which were replaced by a singlet at 1030 cm^{-1} upon dehydration (143). Such behavior, increasing band position after in situ calcination, was similar to that observed for Al_2O_3-supported MoO_3, WO_3, and Re_2O_7. Hence, as was pointed out (143), the distinction between highly dispersed surface phases and otherwise crystalline species may be discerned on the basis of whether hydration–dehydration produces shifts in the terminal metal═O stretching mode. For supported vanadia, however, Raman spectra can appear to be essentially similar to bulk V_2O_5 but without the V_2O_5 having a reducibility that is similar to bulk V_2O_5. This difference is a result of a relatively large Raman scattering cross section for V_2O_5 as compared to the scattering cross section for supported vanadia species. A decrease in reducibility may suggest a significant distortion of V_2O_5 bonding and/or some interaction of a V_2O_5-like species with the support. Hence, Raman spectra of supported V_2O_5 should be closely examined to ascertain whether broadening and peak splittings have occurred.

Raman spectroscopy also has been used to investigate vanadium speciation in catalysts that relates to catalyst deactivation. Such deactivation is, for example, associated with V poisoning of fluid catalytic cracking (FCC) catalysts. Recent novel Raman results in this area are described in the final section of this chapter.

Other Catalysts

Raman spectroscopy has been used to investigate many other types of catalysts. This short section describes some of these Raman investigations and outlines their implications for catalysis.

Hafnia and Zirconia

The selectivity for olefin production by the dehydration of 2-octanol over bulk ZrO_2, HfO_2, and mixed ZrO_2-HfO_2 has been studied (178,179). This selectivity depended on the composition of the mixed catalysts, with over a 50% drop in 1-octene production for samples containing Zr/(Zr + Hf) mole ratios less than approximately 0.8. These HfO_2-ZrO_2 catalysts can be considered to be an ideal metal alloy analog because of the nearly identical ionic size, valence, and oxide structure of Hf and Zr. In addition, ZrO_2-HfO_2 forms a solid solution over the whole range of Zr/(Zr + Hf) mole ratios of $0 \rightarrow 1$, the Raman analysis of which shows gradual shifts in band positions instead of band splittings (180,181). Such monotonic shifts in frequency, in going from pure ZrO_2 $\rightarrow$ mixed ZrO_2-HfO_2 $\rightarrow$ pure HfO_2, was confirmed in the Raman analysis by Simpson and Davis (182). The constant variation in stoichiometry of the solid solution, as monitored by Raman spectroscopy (182) and surface techniques such as ISS and Auger (178,179), has not been able to distinguish structural/chemical entities responsible for the stepwise change in alcohol dehydration selectivity.

Chromia

An investigation of chromia-containing catalysts was summarized by Stencel, Bradley, and Brown (183). These Raman and infrared results showed the advantages of using Raman as opposed to infrared spectroscopy for delineating structural features of supported species. The detection of CrO_4^{2-} units in the NiO-Cr_2O_3/$MgSiO_3$ was affirmed by the similarity in spectra of the surface species and the spectra of standard compounds such as Na_2CrO_4, $MgCrO_4$, and $Al_2(CrO_4)_3$. Iannibello et al. (134,184) had also detected CrO_4^{2-} in alumina-supported chromia. Free CrO_4^{2-}, with T_d symmetry, has two bands at 847 and

884 cm^{-1}, assigned to A and F_2 modes, respectively (185,186). Interaction of this tetrahedral species with the support distorts its symmetry and causes the formation of an asymmetric band in the 820 cm^{-1} region. Cr_2O_7, expected to contain a CrO symmetric stretch near 900 cm^{-1} (187), and CrO_2 or Cr_2O_3, not expected to contain CrO modes in the 750 to 900 cm^{-1} region (188,189), were not observed on the catalyst. Infrared data confirmed this conclusion. Raman peaks at 585 and 545 cm^{-1} suggested the presence of $Cr(OH)_3$. Use of the NiO-$Cr_2O_3/MgSiO_3$ catalyst under methanation conditions caused removal of the 820 cm^{-1} band; this removal was attributed to the reduction of $Cr^{6+} \rightarrow Cr^{3+}$ or $\rightarrow Cr^{2+}$. The $Cr(OH)_3$ was still observed after methanation testing. Carbon was deposited on the catalyst during testing, the Raman spectra of which showed bands at approximately 1600 and 1375 cm^{-1}. Further characterization of these carbon peaks was not performed.

Antimony, Tin, and Titanium Oxides

Recently, Raman spectroscopy was used in conjunction with infrared and Auger spectroscopy and X-ray diffraction to define structural features of Sn-Sb and Sb-Ti oxide catalysts used in propene oxidation (190). The incorporation of only 5 to 10 at% Sb oxide into SnO_2 caused the disappearance of characteristic SnO_2 bands (191), which were located at 775, 630, and 475 cm^{-1}. Broad bands at 735 and 450 cm^{-1} became more intense with increasing Sb oxide concentration, but these positions were not in agreement with those expected for: Sb_2O_3, with bands at 410, 470, and 610 cm^{-1} (192); Sb_2O_5, which has bands at 422, 510, 590, and 817 cm^{-1} (193); or Sb_6O_{13}, which has bands at 420, 470, 550, and 850 cm^{-1} (189). Hence, it was concluded that the surface Sb oxide was amorphous. Additionally, Sb ions were incorporated into the SnO_2 lattice as a solid solution having an ill-defined structure. Only bands attributed to the anatase support were found for Sb oxide–TiO_2 catalysts. These Raman results provided minimal, but important, insight into the structure of these catalysts. It was postulated that the oxidation activity was the result of tin oxide, and that the acrylaldehyde activity was the result of the Sb oxide.

Lanthana and Paladium

The chemical reactions occurring during the preparation of SiO_2- and La_2O_3-supported Pd were discussed relative to Raman vibrational activity by Chan and Bell (194). This type of catalyst had been shown to be active and highly selective for the synthesis of CH_3OH from synthesis gas (195,196). The spectrum of the lanthanum oxide support depended on its degree of hydration, accomplished by refluxing La_2O_3 in boiling H_2O. Band positions were presented for lanthana-derived materials during the decomposition of $La(OH)_3$ to $[La(OH)_2]_2O$ to $LaO(OH)$ to La_2O_3. In addition to these data on the support, the acquisition of in situ data for H_2PdCl_4-impregnated La_2O_3 and SiO_2 suggested that $PdCl_2$ was present in

Table 4.5. Positions of Raman bands in lanthanum hydroxide decomposition (cm^{-1}).

$La(OH)_3$	$[La(OH)_2]_2O$	$LaO(OH)$	La_2O_3
140	122	122	107
	133	131	
227	200	191	195
	227	216	
283	278		
	306	306	
342	339	340	
450	378	378	410
	423	423	
603			

alpha and beta forms on the SiO_2 but not on the La_2O_3 (see Table 4.5). The alpha form of $PdCl_2$ contains Raman bands at 274, 313, and 370 cm^{-1}, whereas the beta form contains bands at 147, 206, and 278 cm^{-1} (197–202). These two forms were observed by Raman spectroscopy and by X-ray diffraction. Calcination at temperatures of 500°C led to the decomposition of Pd chloride, but the formation of PdO was not detected by Raman spectroscopy.

Supported Titania

In another Raman investigation of catalyst preparation, Reichman and Bell (203) examined the interaction of $TiCl_4$ and HCl with SiO_2. In a concentration range of 0.1 to 0.25 *M* titanium, only one Raman band, at 970 cm^{-1}, was observed; it was assigned to the Ti=O stretching mode of TiO^{2+} (204). Two other peaks, at 395 and 910 cm^{-1}, appeared for 0.5 to 1.0 *M* titanium concentrations. These bands were assigned to the symmetric and antisymmetric stretching modes of a Ti-O-Ti dimer of the form $(Ti\text{-}O\text{-}Ti)^{6+}$. Changes in the concentration of HCl caused the formation of Raman bands that were attributed to a Ti oligomer, represented by $[Ti_8O_{12}(H_2O)_{24}]Cl_8 \cdot HCl \cdot 7H_2O$. Decomposition of this species occurred upon its incorporation onto a SiO_2 support. Concomitantly, an amorphous and highly dispersed surface phase developed. High temperature calcination (~550°C) of the amorphous phase was shown to cause development of bands characteristic of anatase.

RAMAN STUDIES OF ADSORBED SPECIES

Introduction

As previously noted, fluorescence associated with laser excitation of Raman spectra from catalysts can cause severe data acquisition problems. Such fluorescence may be insurmountable when molecules at submonolayer levels of adsorption are to be monitored. Hence, early Raman investigations of adsorbed species, during 1965 to 1977, focused on the more tractable amorphous silica and alumina materials in an effort to compare Raman analyses with the wealth of infrared data available for amorphous oxide catalysts. In particular, the catalytically important acid properties of these amorphous supports were studied during the initial Raman experiments.

Adsorption of molecules, utilized to probe the acidic and basic properties of catalysts and catalyst supports, has been studied for over 70 years. These investigations have provided information on the number and strength of acid sites that, for example, are responsible for carbonium ion hydrocarbon transformations such as cracking, alkylation, and isomerization, also, dehydration and dehydrogenation of alcohols are catalyzed by acidic and basic sites, respectively. Spectroscopically, the traditional technique for monitoring adsorption has been infrared spectroscopy. Infrared data not only can provide quantitative information on adsorbates, but also can give the distribution between Lewis and Bronsted acidity. Although the normally intense absorption of the incident infrared radiation by supports or the catalyt itself sometimes has led to insurmountable experimental problems, the low Raman scattering cross sections of such supports were believed to circumvent such sampling-dependent problems. In addition, Raman spectroscopy is completely complementary to infrared spectroscopy and details molecular speciation to the same degree as does infrared.

The low Raman scattering cross section also can be detrimental to the study of adsorbates, especially in the case of adsorption onto single crystal metal supports or otherwise low surface area supports. However, recent advances in experimental equipment and in theoretical understanding of the symmetry selection rules for Raman activity of surface-adsorbed species have provided a new impetus for monitoring adsorption by Raman spectroscopy. This section presents such Raman studies of planar surface adsorption in conjunction with Raman results of adsorption on high surface area supports.

High Surface Area Supports

The first reported studies of Raman data from molecules adsorbed on high surface area supports utilized mecury arc lamp excitation of aromatic hydrocarbons (205–208). The use of such an excitation source extremely limited the experimental possibilities, but was necessary before the advent of lasers. Subsequent to laser development, Hendra and Loader (209,210) were the first investigators to utilize a He-Ne laser for Raman spectral excitation from adsorbed molecules. A sensitivity of approximately 0.1 monolayer was demonstrated for bromine, trans-dichloroethylene, carbon tetrachloride, and acetaldehyde on silica gel. The symmetry of these molecules was optimal for obtaining strong Raman signals even though the relatively low frequency of the He-Ne irradiation produced a substantially inefficient Raman scattering across section as a result of the ν^4 dependence of such scattering.

In 1970 and 1971, Hendra and Loader published Raman spectra of CCl_4, Br_2, CS_2, acetonitrile, and other adsorbates on silica and silica-aluminas (211–213). The possibility that the resulting spectra originated from a gaseous or liquid adsorbate was discounted because of shifts in band positions relative to the known band position in neat gaseous or liquid samples and because of the appearance and/or disappearance of some adsorbate vibrational modes. These data and the discussions they produced were the first attempt to describe the effects of chemisorption versus physadsorption on characteristic Raman vibrational modes.

Hendra et al. (214,215) continued such studies with the analysis of pyridine chemisorbed onto inorganic oxides. This effort paralleled contemporary infrared studies of pyridine chemisorption and the identification of Lewis and Bronsted acidity in oxidic catalysts (216–221). The bands that were closely monitored included the skeletal vibrations of pyridine near 1000 cm^{-1} and the CH stretching vibrations near 3000 cm^{-1}. Spectra of adsorbed pyridine, including their peak widths, positions, and relative intensities, were compared to spectra of liquid pyridine. The degree of interaction between pyridine and the oxide surfaces was shown to be defined by the vibrational position of Lewis coordinated species; for example, liquid pyridine had the C-N stretching vibration at 991 cm^{-1}, whereas on titania it was at 1016 cm^{-1}, on alumina at 1019 cm^{-1}, on silica-alumina at 1020 cm^{-1}, and on chlorided alumina at 1023 cm^{-1}. This pioneering Raman study could not identify bands that were distinctive of Bronsted acidity, although pyridine "physadsorbed" or H-bonded to the oxide surfaces was delineated. Additionally, the significant outcome of these studies, as also suggested by the authors, was that Raman spectroscopy was rapid, could monitor significant changes in vibrational characteristics of adsorbates that were associated with sorbent acidity, and was experimentally compatible with equipment needed to acquire adsorption isotherm data from the same sample.

Lewis acidity on Al_2O_3 also was studied using Raman spectroscopy (219); early Raman studies of adsorption on Al_2O_3 included nitrobenzene (220,221), olefins, and the nitriles CH_3CN, C_2H_5CN, and C_6H_5CN (222). The reaction of nitriles and olefins over Al_2O_3 at 25°C probably was the first Raman spectroscopic investigation showing changes in surface species with time of reaction.

Buechler and Turkevich (5) in 1972 published Raman spectra of porous Vycor, and investigated, to some extent, the spectral characteristics of various adsorbates. Because porous Vycor was investigated in early infrared studies (223,224), the corresponding Raman data on adsorbates were important for comparison of the capabilities of these vibrational techniques. For example, adsorption of unsaturated and halogenated molecules (225,226), benzene (5,223–227), and thiophene (228) produced significant differences in Raman spectra as compared to the corresponding neat liquid spectra. Such data showed that Raman spectroscopy could be utilized in providing structural information on adsorbates.

Bahnick and Person (229) described intensity increases of nearly a factor of two for some vibrational modes of CCl_4 when it was placed into solution with electron donor molecules. These increases, ascribed to charge transfer interactions, were part of the experimental evidence leading to further investigations in which the possibility of previously unknown vibrational band intensity increases of adsorbates were examined (230). Saturated hydrocarbons and hydrocarbons having a center of symmetry (e.g., tetrachloroethylene and benzene, respectively) were adsorbed on porous silica surfaces, and their vibrational bands were monitored by Raman spectroscopy. Table 4.6 displays some of the results (230). In all cases, small changes (1–3 cm^{-1}) in band positions away from the

Table 4.6. Positions and relative intensities of liquids and after adsorption onto silica (230).

	LIQUID		ADSORBED		
COMPOUND	$\nu(cm^{-1})$	I_{rel}	$\nu(cm^{-1})$	I_{rel}	ASSIGNMENT
Benzene	609	0.02	607	0.11	ν_{18} (E_{2g})
	990	1.00	991	1.00	ν_2 (A_{1g})
	1178	0.08	1176	0.21	ν_{17} (E_{2g})
	1587	0.08	1586	0.16	ν_{16} (E_{2g})
	1606	0.06	1607	0.13	$\nu_2 + \nu_{18}$ (E_{2g})
	2947	0.12	2947	0.14	$2\nu_{13}$ (A_{1g},A_{2g})
	3060	1.48	3061	2.53	ν_1 (A_{1g})
CCl_4	216	0.37	218	0.68	ν_2 (E)
	313	0.37	313	0.71	ν_4 (F_2)
	453, 456, 460	1.00	460	1.00	ν_1 (A_1)
	759	0.34	760	0.38	ν_3 (F_2),
	789	0.38	790	0.61	$\nu_1 + \nu_4$ (F_2)
Tetrachloroethylene	238	0.45	239	0.62	ν_3 (A_g)
	350	0.13	350	0.27	ν_6 (B_{1g})
	448	1.00	450	1.00	ν_2 (A_g)
	512	0.10	513	0.06	ν_8 (B_{2g})

corresponding positions in neat samples were measured; relative band intensities were changed dramatically upon adsorption. In centrosymmetric molecules, infrared and Raman activity are mutually exclusive. For the molecules listed in Table 4.6, normally infrared active bands were not observed in Raman spectra upon their adsorption onto SiO_2 surfaces. Hence, the adsorbate–adsorbent interactions on SiO_2 were not sufficiently perturbing to cause severe distortion of the molecular symmetry, and, as a result, to cause otherwise Raman inactive modes to appear. However, the anisotropic electric field experienced by the adsorbate on a surface was described as the important factor influencing relative band intensities.

The Raman study by Angell (231), who investigated adsorption of carbon dioxide, propylene, acetonitrile, and acrolein on zeolites A, X, and Y, was performed in conjunction with an analysis of the characteristic vibrational spectra of these zeolites. The data for the adsorbed molecules were similar to the Raman data for the liquid state of the molecules. Only in the cases of the C═C stretching frequency of propylene and the C≡N stretching frequency of acetonitrile were significant differences found, which were indicative of the effects of adsorption.

The adsorption of acetylene and dimethylacetylene on A and X zeolites was monitored with Raman and infrared spectroscopy by Tam, Cooney, and Curthoys (232,233). The five fundamental vibrational modes of C_2H_2 (point group $D_{\infty h}$, centrosymmetric—see reference 234) were observed in its adsorbed state. The frequencies of these modes were significantly decreased as compared to their gaseous or liquid frequencies. An infrared forbidden band at ca. 1950 cm^{-1} appeared in the spectra of the adsorbed state of C_2H_2, indicating that the activity of this ν_2 vibration was the result of a lowering of molecular symmetry. Such a symmetry change upon adsorption was theoretically modeled using normal coordinate analysis and a C_{2v} symmetry. In this symmetry the C_2H_2 still has linear C≡C-H bonds. Very good agreement between calculated and experimental vibrational frequencies were obtained, where the Raman active modes on NaA had values of ν_1(C-H): 3321 cm^{-1}, ν_2(C≡C): 1953 cm^{-1}, and ν_4-δ [(H-C≡C)]: 642 cm^{-1}. The research by Tam et al. (232,233) also showed that the type of cation (e.g., Li, Na, K, Cs, Mg, Ca, or Ba) dramatically influenced the frequency of the ν_2(C≡C) symmetric stretching mode for adsorbed C_2H_2.

A similar effect of cation-induced frequency shifts on the symmetric ring breathing mode for pyridine/13X was observed (235). This study related directly to previous work on the adsorption of pyridine on Y zeolites (236), and on the polarizing effects of the cation (236,237). Here, a linear correlation between the ν_1 frequency of the symmetric ring breathing mode (i.e., where the six-membered aromatic ring expands and contracts in a uniform motion) and the electostatic potential of the cation–framework pair was demonstrated. Hardin, Klemes, and Morrow (235) used Y zeolite containing the cations NH_4^+, CS^+, K^+, N^+, Li^+, Pb^{2+}, Cd^{2+}, Zn^{2+}, Co^{2+}, and Ni^{2+}, and were successful in defining the dis-

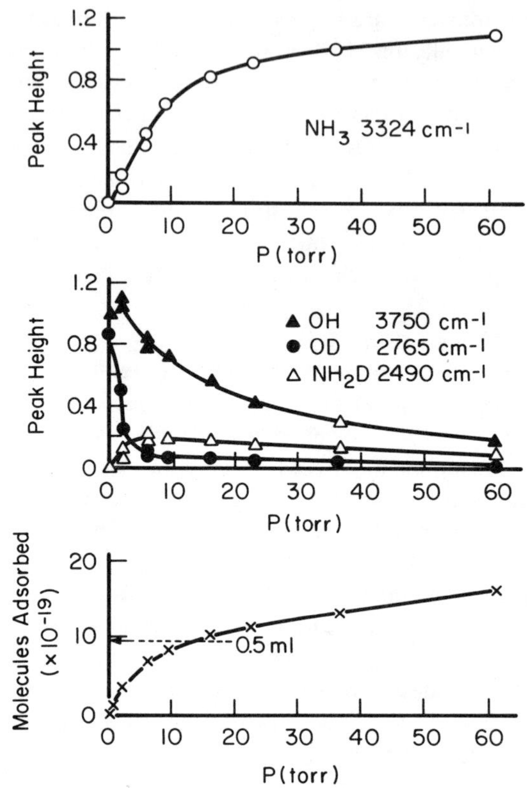

Figure 4.18. Peak heights normalized to those of the 800 cm^{-1} bulk band as a function of pressure (upper graphs) and adsorption isotherm (lower graph) for the partially deuterated surface. The top points of those marked with double symbols represent the values for the second scan of the spectrum (about 7 hours later) at the same pressure (240).

tribution of these cations among the S_I and S_{II} sites. Frequency decreases in ν_1, intensity changes of ν_1 upon adsorption, and the Raman analyses of model solutions, (i.e., metal nitrates in a solution of pyridine in nitromethane) showed that the intensity of the ν_1 band could be used as an "adsorption isotherm," as previously described (238). In addition, the charge-to-radius ratio of the cations was shown to control pyridine–zeolite interactions in X and Y zeolites (239).

Murray and Greytak (240) investigated the combined perturbational effects of NH_3 adsorption on the vibrational modes of the adsorbate and the adsorbent. In this study, NH_3, adsorbed onto porous Vycor at ~150°C, could be detected at as small as 0.01 monolayer coverage. The hydroxyl-covered Vycor surface produced a Raman active Si-O stretching mode at 980 cm^{-1} and a corresponding strong OH stretching mode at 3750 cm^{-1}. Upon NH_3 adsorption, as indicated in

Figure 4.18, the intensity of the 3750 cm^{-1} band decreased linearly, with a corresponding increase in the NH stretching modes at 3414, 3324, and 3285 cm^{-1}. Notably absent from these data was the highly intense and broad band centered near 3100 cm^{-1}, which normally occurs in infrared spectra of NH_3 adsorption on SiO_2 surfaces. Such an infrared band, caused by H-bonded surface hydroxyl groups (241), can decrease the quantitative measurement of surface coverage by infrared analyses. An estimate of the concentration of surface OH groups by Raman spectroscopy was 7.1×10^{20} OH/cm^3 or 2.2 OH/100 Å^2, a value in close agreement with other literature data (223,224,242).

The effect of NH_3 adsorption on the Si-O vibrational mode at 980 cm^{-1} was not as dramatic as that observed in Figure 4.18. Nevertheless, a gradual shifting of the 980 cm^{-1} band to 998 cm^{-1} was observed as a result of the increased polarity of the surface SiO-OH bond, possibly during donation of a lone pair electron to the nitrogen within NH_3. Similarly, near a one-monolayer coverage the H-bonded NH_3 exhibited band shapes and positions close to those for aqueous NH_3 solution. This correspondence was considered to indicate a condensed phase having OH-N bonding that was experiencing perturbational effects of nearby NH_3. In addition, chemisorbed NH_3, resulting from NH_3 coordinately bonded onto impurities on the surface or onto electron-deficient Si sites was observed in this study (240). The concentration of these sites, ~0.5 site/100 Å^2, was in near agreement with infrared data of Peri (243). Murray and Greytak (240) also used deuterated Vycor and observed H–D exchange between the surface and the adsorbed NH_3. This rapid reaction, that is, $NH_3 + OD \rightarrow NH_2D + OH^-$, was not, however, analyzed in detail.

The adsorption of NH_3 was investigated by Zhang and Schrader for the purpose of delineating the NH_3 decomposition reaction over Fe catalysts (244). Advantages cited for the use of Raman spectroscopy as compared to infrared spectroscopy were that the Fe catalysts cause large absorption of the infrared beam, whereas the Raman scattering cross section of Fe-O or other Fe-associated bonds is very small; simultaneously, the Fe-N, N-N, and N-H bonds have rather strong scattering cross sections. Assignments from this NH_3/Fe study are listed in Table 4.7. Notably, some of the spectra that were reported were obtained with the Fe catalyst at 400°C. This experiment, then, was truly in situ; the N-N stretching band near 1940 cm^{-1} was assigned to a "flat-lying, double-end-on bridge-type" species. This species was believed to be in equilibrium with adsorbed N_2 having an end-on configuration. The Raman data suggested that the decomposition reaction could be described as:

$$NH_3 \rightarrow NH_{3(a)} \rightarrow H_{(a)} + NH_{2(a)} \rightarrow 2H_{(a)} + NH_{(a)} \rightarrow 3H_{(a)} + N_{(a)}$$

These data and peak assignments pointed to new information not previously obtainable with the use of infrared or EELS instrumentation.

Table 4.7. Raman assignments for adsorbed species associated with NH_3 synthesis and decomposition.

BAND POSITION (CM^{-1})	ASSIGNMENT
1951, 1902	Fe—H from $H_{(a)}$
1940, 204	N≡N from $N_{2(a)}$
1090	Fe≡N from $N_{(a)}$
895	Fe═N from $NH_{(a)}$
500	Fe—N from $NH_{2(a)}$
465, 443, 410	Fe—N from $N_{2(a)}$
140	Fe—N from $NH_{3(a)}$

Schrader and Cheng (245) analyzed pyridine adsorption on gamma-Al_2O_3, Mo/Al_2O_3, and Co-Mo/Al_2O_3 as a function of Co and Mo loading. They were successful in defining changes in Lewis acidity as a function of loading of the active, supported phase, and in determining how these changes related to the surface structure of the supported phase. They also observed creation of Bronsted acid sites at loadings of Mo less than theoretical monolayer levels. For example, Table 4.8 lists the band positions of pyridine adsorbed on Mo/Al_2O_3 and Co-

Table 4.8. Raman band positions of pyridine adsorbed on Mo/Al_2O_3 and Co-Mo/Al_2O_3 (245).

		ν (cm^{-1})			
SAMPLE (LOADING)	LIQ		ADSORBED SPECIES		LIQ
δ-Al_2O_3	991	998		1017	1032
Mo(1.25)*	991	1000		1017	1032
Mo(2.5)	991	1002	1006	1017	1032
Mo(5)	991	1004	1008	1017	1032
Mo(7.5)	991	1004	1008	1017	1032
Mo(10)	991	1004	1008	1017	1032
Mo(15)			1008	1017	
Co(3)Mo(1.25)				1014w	1032
Co(3)Mo(2.5)	991			1014	1032w
Co(3)Mo(5)	991w			1014	1032
Co(3)Mo(10)	991			1014	1032
Co(3)Mo(15)				1014br	1032
Co(0.5)Mo(7.5)	991			1016	1032
Co(1)Mo(7.5)	991			1016	1032
Co(2)Mo(7.5)	991			1014	1032
Co(3)Mo(7.5)	991w			1014	1032w

*Mo loading expressed as MoO_3, Co loading expressed as CoO (pyridine pressure = 5 Torr), w = weak, br = broad.

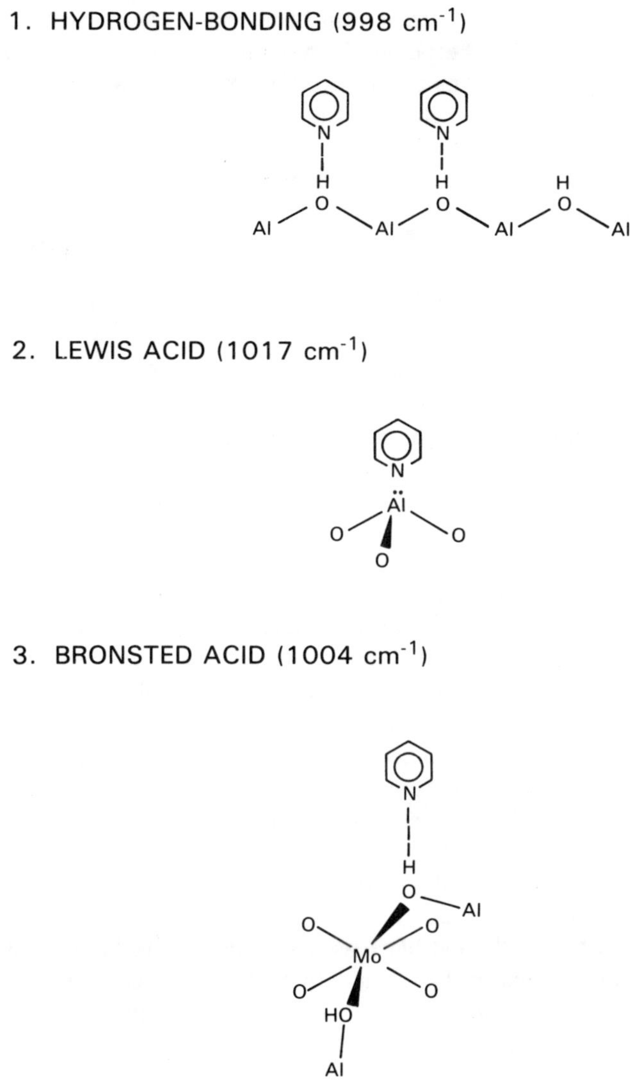

Figure 4.19. Models for hydrogen bonding, Lewis acid, and Bronsted acid sites on γ-Al_2O_3 (245). (Reprinted with permission from *The Journal of Physical Chemistry*, Copyright 1983, The American Chemical Society.)

Mo/Al_2O_3. The Lewis coordinated pyridine, as depicted in Figure 4.19, was associated with the 1017 cm^{-1} band. Adsorption with molybdena loadings in the range 5 to 10% also produced Bronsted associated pyridine at 1004 cm^{-1} (see Figure 4.19). With the addition of Co, bands associated with Bronsted sites were no longer observed. Although bulk, crystalline MoO_3 was observed at higher

Mo loadings, no systematic decrease in Lewis acidity was observed as the Mo loading increased up to the point of crystalline MoO_3 detection. Hence, the extent of formation of a Mo monolayer could be described, whereas, in agreement with Mone (246), the interaction of Co with the Al_2O_3 surface and with impregnated Mo also could be discerned.

Segawa and Hall (247), at about the same time as Schrader and Cheng (245), published an infrared and TGA study of pyridine adsorption on Al_2O_3 and Mo/Al_2O_3. Their infrared data on the oxidic form of Mo/Al_2O_3 showed, in agreement with Schrader and Cheng, that Lewis and Bronsted acid sites were present on Mo/Al_2O_3. In addition, Segawa and Hall (247) detected these sites on Al_2O_3-supported chromia, rhenia, and tungsta catalysts. However, upon H_2 treatment at 500°C, Bronsted acid sites on the Mo/Al_2O_3 catalysts were eliminated. Indeed, the information presented by Segawa and Hall dictates a need for more in-depth analysis of the role of such acidity on these catalysts.

More recently, pyridine adsorption on Mo/Al_2O_3 and Mo/SiO_2 was discussed in relation to corresponding X-ray diffraction data obtained from the same catalysts (248). On SiO_2 the detection of crystalline MoO_3 occurred near 8 mol% molybdena, but its appearance did not change the amount of pyridine that was adsorbed onto the surface, contrary to the data of Schrader and Cheng (245). Additionally, Bronsted acid sites were detected by their characteristic frequencies from the pyridinium ion at 1010 and 1030 cm^{-1}, whereas H-bonded pyridine was assigned to 1000 and 1032 cm^{-1} bands that were also observed. Although SiO_2 surfaces possess only Bronsted acidity, the detection of crystalline MoO_3 at an 8% molybdena loading coincided with the observation of Lewis coordinated pyridine having a band at 1020 cm^{-1}. Hence, molybdena on Al_2O_3 was described as producing Bronsted acid sites (245), whereas molybdena on SiO_2 was said to produce Lewis acidity (248).

For Mo/Al_2O_3, Ismail et al. (248) did not observe crystalline MoO_3 even at loadings near theoretical monolayer levels. Lewis acidity, as determined by the presence of a 1020 cm^{-1} band, decreased with increasing Mo content such that at 16 mol% molybdena no Lewis acidity could be detected. This finding was thought to indicate the substitution of surface AlO_6 groups by MoO_6 octahedral groups.

Adsorption on Metal Surfaces

The use of Raman spectroscopy to investigate adsorption on high surface area Raney Ni and Ni/SiO_2 was first reported in 1977 (249). This effort successfully used a rotating sample holder, originally designed by Kiefer and Berstein (250,251), to avoid localized heating and degradation of the highly absorbing samples. Carbon monoxide is not a strong Raman scatterer; even so, spectra of adsorbed molecules could be obtained for a CO pressure of less than 50 Torr. Stretching vibrations of chemisorbed CO were found at 1903, 1940, and 2100 cm^{-1}, with the 1940 cm^{-1} band containing fine structure at 1945 and 1936 cm^{-1}. The 2100 cm^{-1} band was assigned to CO stretching of a Ni-CO species, in agreement with the Eischens infrared assignment (252), whereas the 1940 cm^{-1} bands were associated with bridged, $Ni(CO)_2$ species. In addition the Ni-C stretch at ~380 cm^{-1} and bending δ(Ni-C═O) modes near 460 cm^{-1} were observed.

Krasser, Fadini and Renouprez (253) used extended Huckel calculations to provide insight into changes of vibrational frequencies as a function of coverage of CO on silica-supported Ni. Using linear and bridge bonded species, with five Ni atoms involved in the linear bonded case and four Ni atoms involved in the bridge bonded case, they were able to discuss the expected symmetry of CO on the Ni surface. A C_{4v} symmetry for linear bonded CO led to the expectation of three internal vibrational modes for Ni-CO: (1) a symmetric ν (CO) stretching vibration of A_1 symmetry, (2) a symmetric ν (Ni-C) stretching vibration having A_1 symmetry, and (3) a degenerate deformation δ (Ni-C-O) having E symmetry. In the bridge bonded case, in which the site symmetry was also C_{4v}, eight vibrational modes were expected, containing symmetric (A_1), antisymmetric (B_1), and degenerate (E) Ni-C stretching vibrations, a Ni-CO deformation having A_1 symmetry, a antisymmetric B_2 and degenerate E deformation of Ni_2-C bonds, and an antisymmetric B_1 and degenerate E deformation of Ni-C-O bonds. The Raman spectra were of sufficient quality to enable a precise plot of a 140 cm^{-1} decrease of the C-O stretching vibration for bridge bonded species as the pressure

Table 4.9. CO vibrations on Ni(100) (Reprinted from Ref. 256 by courtesy of Marcel Dekker, Inc)

SPECIES	ASSIGNMENT		
Linear site (C_{4v})			
A_1	ν_1(C—O)	2088	2050
	ν_2(Ni—C)	437	
E	ν_3 δ (OCNi)	411	419
	ν_4 δ (CNiNi)	82	
Bridge site (C_{2v})			
A_1	ν_1 (C—O)	2099	2028
	ν_2 (Ni—C)	457	466
B_1	ν_3 (Ni—C)	649	
	ν_4 (OCNi)	184	
B_2	ν_5 δ ($OCNi_2$)	572	
	ν_6 δ (CNiNi)	76	

of CO was decreased from near atmosphere down to 0.05 Torr. Normal coordinate analyses enabled the calculation of force constants and an assignment of bands that were observed. Force constants for the C-O and Ni-C bonds were close to those reported for $Ni(CO)_4$

Stencel and Bradley obtained Raman data for the adsorption of CO on smooth, oriented surfaces of Ni (254,255). This work, and further work by Bradley and coworkers (256–259), led to the assignments of CO modes on Ni(100), as shown in Table 4.9. These Raman data are in fair agreement with the interpretations of Richardson and Bradshaw (260), but are in some disagreement with results from EELS experimentation (261) and from Raney Ni experiments (249,253).

By 1984, improvements in sampling geometry and data processing and the advent of multichannel detectors on Raman instrumentation allowed extremely sensitive measurements of CO adsorbed on Ni(111), in addition to the detection of δ(CNiNi) and δ(OCNi) vibrations at 80 and 185 cm^{-1}, respectively (256). Figure 4.20 shows the type of species believed to be present and the corresponding vibrational frequencies. These assignments were the culmination of carefully controlled CO adsorption experiments using Ni(111), accompanied by polarization measurements and arguments based on known CO bonding. Coadsorption of H_2 led to a 23 cm^{-1} decrease in the bridge-bonded CO stretching frequency.

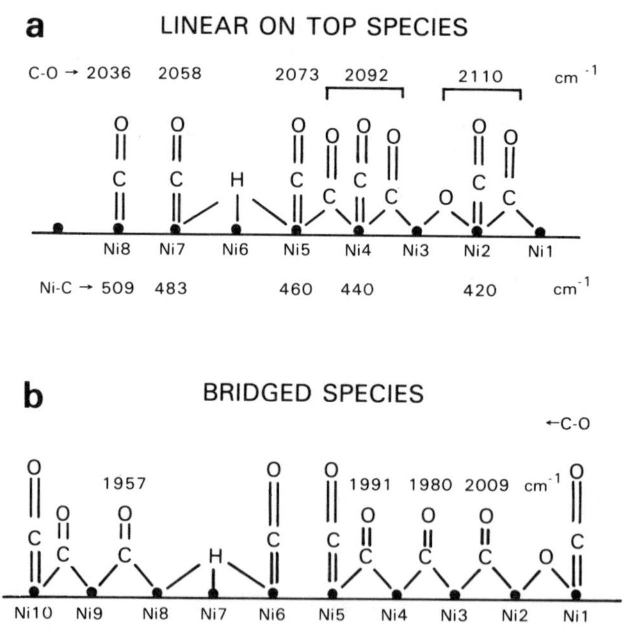

Figure 4.20. (a) Schematic of a proposed linear on-top CO species coadsorbed on a Ni surface with H and O. (b) Schematic of a proposed bridged CO species coadsorbed on the Ni surface with H and O (257).

Such a decrease had not been previously observed in adsorption experiments that used EELS (262), probably because of the relatively poor resolution of EELS as compared to Raman spectroscopy. Other noteworthy findings of this Raman study included adsorbed O_2 species and the formation of hydrides.

Following their success in observing CO/Ni(100) and CO/Ni(111), Bradley and coworkers (263) were the first investigators to study vibrational spectra of CO adsorption on planar polycrystalline Co. Figure 4.21 illustrates their interpretation of the modes that were observed. Comparison of these data with those in Figure 4.20 shows that generally lower vibrational frequencies are observed for CO on Co than for CO on Ni. This would imply stronger Co-C bonding than Ni-C bonding. Heats of adsorption of CO/Ni are less than of those of CO/Co (264).

The observation of chemisorbed molecules on smooth, planar surfaces by use of normal Raman spectroscopy relies on the maximization of a number of experimental variables. Optimization of the design and geometry of the adsorption cell and its optics is imperative for detection of such adsorbed molecules. Part of the reason for the successful observation of the Raman spectra of CO/Ni(100), CO/Ni(111), and CO/Co was the careful design of the cell and optics used. The advent of high transmission spectrometers, lasers, computers, and multichannel detectors also has been extremely important. The optical arrangement, a near 60° angle of incidence of the laser beam with respect to the surface, optimized collection of the Raman scattered radiation (257–259,265); and the use of a special cell that had heating and cooling capabilities and simultaneously was usable with surface science techniques such as XPS, LEED, and so on, also was very important in obtaining acceptable signal-to-noise ratios for adsorption on planar surfaces.

Theoretical considerations about radiation emitted by oriented dipoles adsorbed on surfaces, which facilitated investigators' understanding of the maximization of experimental parameters in Raman scattering, were first described by Greenler and Slager (266). Campion and coworkers (267–272) extended this theoretical base by including the contribution of a distribution of observation angles sampled by the input optics of the Raman spectrometer. Hence, for the orientations of a radiating dipole that are shown in Figure 4.22a, the angular distribution of radiation emitted would be as depicted in Figure 4.22b (272–274). By performing angle-dependent Raman studies, in which the scattering angle was varied while the incident angle and polarization remained constant, a determination of the symmetry and orientation of molecules adsorbed on smooth metal surfaces was possible. In addition, producing this type of information on chemisorbed species as a function of the polarizability tensor components (which were discussed in Chapter 2) made possible the determination of changes in electronic density of the chemisorbed species resulting from chemisorption.

The angle-dependent Raman equipment of Campion et al. (272–274) also permitted LEED and Auger measurements on the same sample surface so that

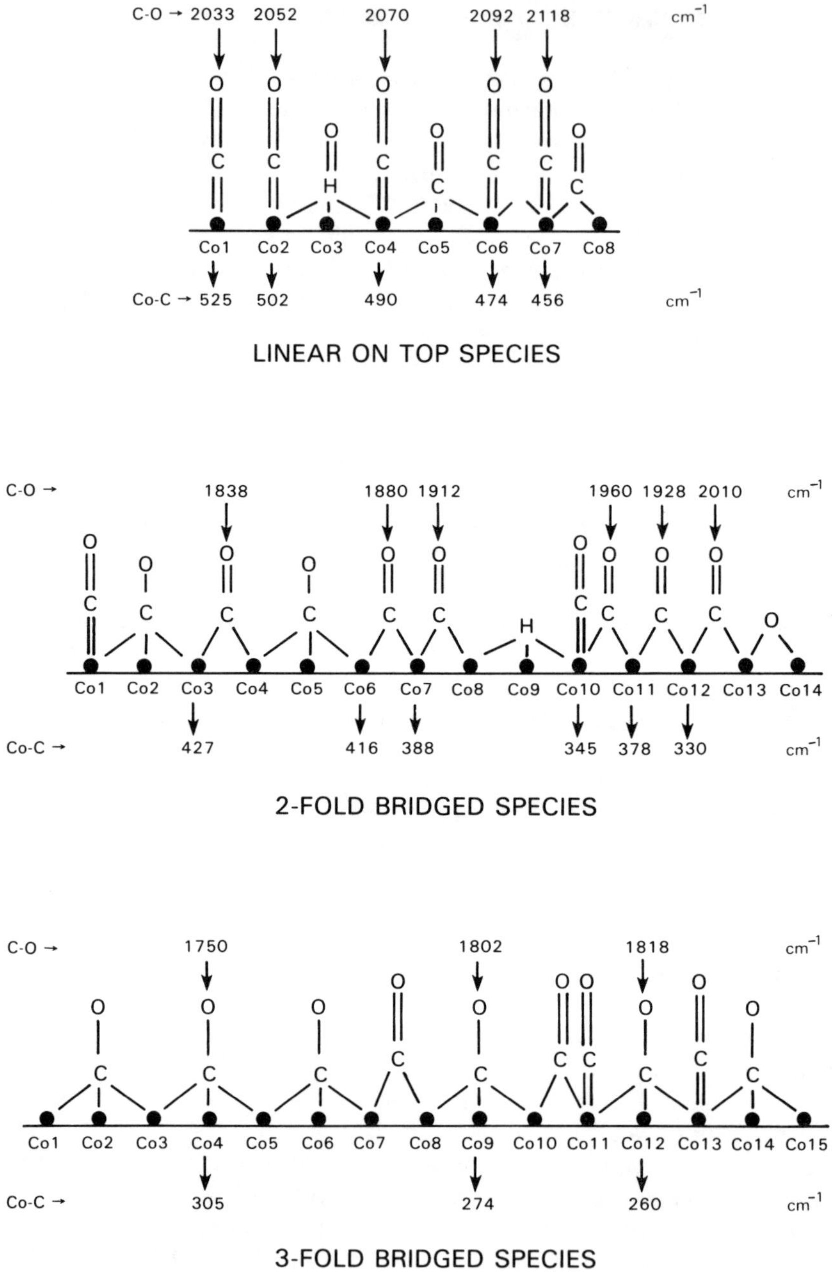

Figure 4.21. Sketches showing proposed models for: the linear on-top CO species coadsorbed on a Co surface with H and O; the 2-fold bridged CO species coadsorbed on Co with H and O; and the 3-fold bridged CO species coadsorbed on a Co surface with different configurations of other CO species. (Reprinted from reference 263 by courtesy of Marcel Dekker, Inc.)

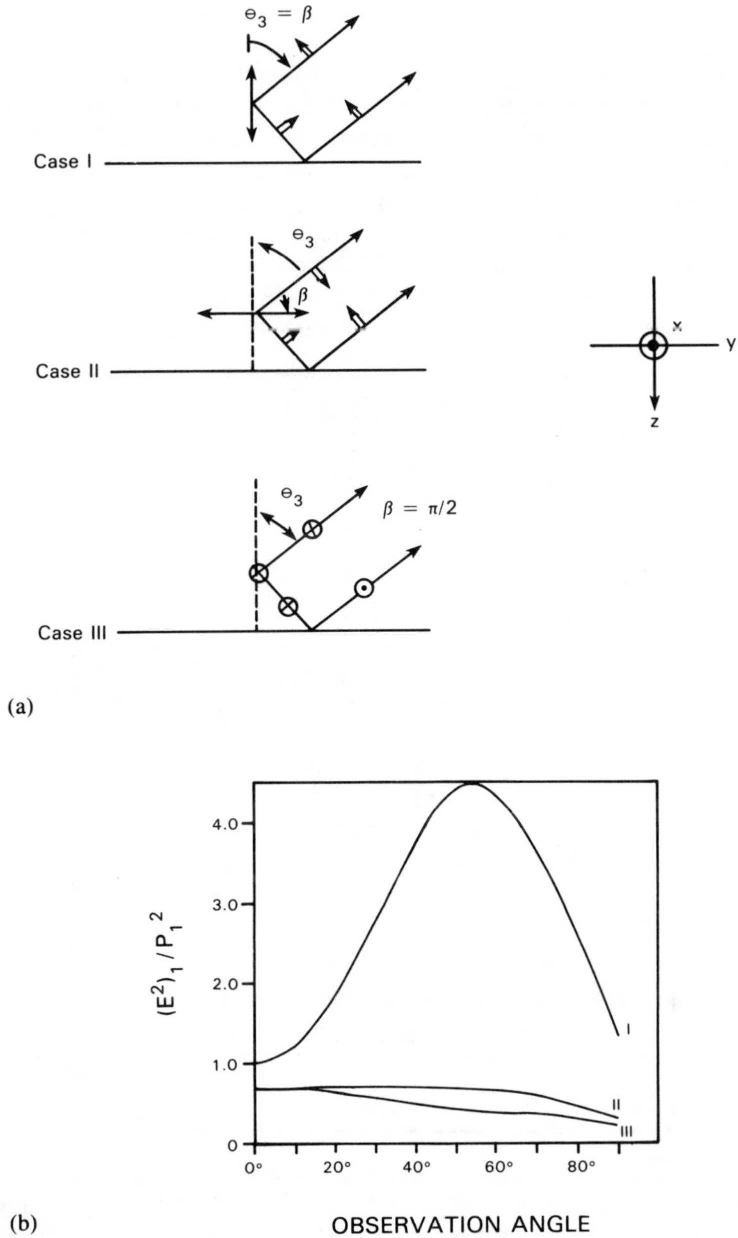

Figure 4.22. (a) Principal orientations of a radiating dipole near a metal surface, where z is the surface normal and yz is in the plane of observation. (b) Angular distribution of radiation emitted by a dipole near a metal surface, for each of the principal orientations pictured above (271).

the amount of adsorbate could be quantified. To show the utility and possibilities of such sophisticated Raman instrumentation, the adsorption of nitrobenzene on Ni(111) was investigated (271,272). Interestingly, a distinct difference existed between spectra at submonolayer and multilayer coverages, as shown in Figure 4.23. Notably, the symmetric NO_2 stretching mode at ca. 1340 cm^{-1} in the multilayer coverage spectrum, a position that is in agreement with liquid nitrobenzene, is completely absent in the ~0.5 monolayer spectrum. Indeed, band positions in the multilayer spectrum at 1580 and 1470 cm^{-1} (C=C stretch), 1515

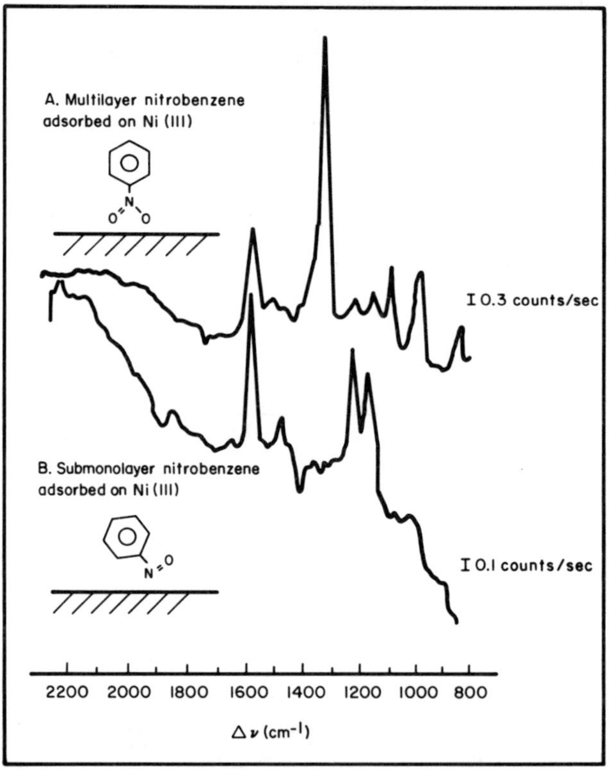

Figure 4.23. (a) Raman spectrum of a very thin film (~ 5 nm) of nitrobenzene condensed on a Ni(111) surface at 100°K; 150 mW of 514.5 nm laser power was used and the spectrum was integrated for 500 seconds. (b) Normal Raman spectrum of less than a monolayer of nitrobenzene adsorbed of Ni(111) at 100°K. The nitrosobenzene resulted from dissociative chemisorption of nitrobenzene. Conditions in (b) were the same as in (a) except that the spectrum was integrated for 1000 seconds (272). (Reprinted, with permission, from the *Annual Review of Physical Chemistry,* Volume *36,* © 1985 by Annual Reviews, Inc.).

and 1340 cm^{-1} (NO_2 stretch), 1160 cm^{-1} (CH in-plane bend), 1100 cm^{-1} (C-N stretch), and 990 cm^{-1} (ring breathing mode) are all similar to those for liquid nitrobenzene. The complete change in band positions for submonolayer coverage, as compared to higher coverages, was attributed to dissociative chemisorption that produced nitrosobenzene and surface NiO. If so, the N=O stretching mode, observed at 1523 cm^{-1} in the gas phase and now assigned to the 1215 cm^{-1} band in Figure 4.23b, decreased by 308 cm^{-1} upon coordination to the Ni surface. This decrease, caused by possible electron transfer into pi* orbitals of the N=O group, is reminiscent of the decrease in CO vibrational frequencies observed in infrared spectra during adsorption onto Ni surfaces.

It must be mentioned that the above studies were performed without the enhancement in Raman intensities that accompanies the SERS effect (See Chapter 5). Even for CO on Cu, a substrate that has been shown to produce the SERS effect, enhancement of Raman intensities upon CO adsorption has been estimated to be only about one order of magnitude (275), rather than the six orders of magnitude increase observed for pyridine on Cu. Chou et al. (276) confirmed the results of Campion (271,272) on the adsorption of nitrobenzene, and investigated the possibility of observing Raman spectra from a monolayer of pyridine on Ni(111). The Chou et al. study also was directed at describing the possibility of enhancements in the pyridine/Ni vibrational spectra, as had been observed for pyridine/Ag (277). However, their Raman data showed no enhancements for the symmetric stretch, ring breathing mode. As a result, a quenching effect on such modes due to the Ni substrate was described; this quenching could be the result of (1) an orientational effect in which only specifically oriented pyridine produced enhancements, or (2) a modification of the polarizability of the adsorbed pyridine such that low scattering cross sections ensued.

Generally, the SERS effect can been identified by the enhancement of band intensities when pyridine is adsorbed on a metal (278–280). Recently, Parker, Siedle, and Hexter (281) showed that the spectrum of pyridine on rhodium colloidal particles exhibited possible SERS effects. The Raman spectra were qualitatively similar to those observed for pyridine on silver (282), where an intense band appeared at 1025 cm^{-1} upon adsorption of pyridine. Hence, a significant enhancement of vibrational intensities of adsorbed species is possible for metal substrates other than just Ag, Cu, and Au.

In another study using Rh as a substrate, in this case Rh supported on Al_2O_3 (283,284), Parker et al. discussed the adsorption of C_2H_2. Here, it was found, for 2.6-nm particles with an approximate 10% Rh/Al_2O_3, that the acetylenic species adsorbed with its symmetry axis essentially parallel to the Rh surface. A representation of this adsorbed species is given in Figure 4.24, where it can be observed that the C≡C-H bonds were no longer thought to be linear (the exact bond angle was not presented). The site symmetry of the adsorbed species was either C_s or C_1, an estimate obtained from arguments based on surface

Figure 4.24. Representation of acetylene bound to a 3-fold site (284). (Reprinted with permission from *Journal of the American Chemical Society*. Copyright 1985 American Chemical Society).

selection rules for Raman spectroscopy (285). The species that was observed was tightly bound, as evacuation of the Raman cell to 10^{-7} Torr did not decrease adsorbate intensity. In addition, the surface species was described to contain a C═C bond order of slightly less than 2. This approximation was a result of observing C═C stretching frequency near 1478 cm^{-1}, a position that is slightly less than that observed for ethylene. The adsorbed species was also very chemically reactive, as the introduction of H_2 caused its immediate disappearance. This surface species on the small Rh particles was discussed as being substantially different from the corresponding species observed when acetylene was adsorbed onto single-crystal Rh.

Raman spectra of H_2 and CO adsorbed onto reduced and carbided fused iron surfaces were presented by Zhang and Schrader (286). For the reduced samples, bands at 1902 and 1951 cm^{-1} were attributed to Fe-H terminal stretching modes, whereas a band at 1625 cm^{-1} was assigned to Fe-H stretching of bridge-bonded H species. Raman spectra of 200°C, CO/H_2 treated catalysts contained the bands listed in Table 4.10. Because iron catalysts under "equilibrium" Fischer-Tropsch (FT) conditions would be expected to contain Fe^0, Fe^{2+}, and Fe^{3+} species

Table 4.10. Raman band frequencies (cm^{-1}) and assignments of adsorbed species on fused iron catalysts. (286)

POSITION (CM^{-1})	ASSIGNMENT
1902, (1948), 1951	Fe—H stretch, terminal H species
1625	Fe—H stretch, bridging H species
1970	C—O stretch, CO/Fe^0 sites
2070	C—O stretch, CO/Fe^{2+} sites
2130, 2160, 2175	C—O stretch, CO/Fe^{3+} sites
1820, 1850, 1890	C—O stretch, CO/Fe^0 multinuclear clusters
1556, 1586, 1600	Vibrations of formate-like species
360, 400, 450	Fe—C stretch of adsorbed CO species

(287,288), and in agreement with infrared measurements, the Raman data in Table 4.10 were said to represent CO bonded to various oxidized states of iron. In addition, the possible importance of Fe^{2+} or Fe^{3+} sites in the formation of alcohols during FT synthesis was discussed. Raman bands in the 1560 to 1600 cm^{-1} region were suggested to infer the presence of CH_xO_y intermediates; the importance of these intermediates in alcohol synthesis was discussed.

ZEOLITES—SYNTHESIS, STRUCTURE, AND DEACTIVATION

Introduction

Zeolites are used extensively for their catalytic and adsorption properties. For example, the fluid catalytic cracking (FCC) market for zeolites in 1990 alone will have an estimated value of $275,000,000 in U.S. dollars and will involve a total tonnage of approximately 405,000,000 tons (289). The types of zeolite applications include catalytic cracking, isomerization, alkylation, carbonylation, polymerization, aromatization, and dehydrogenation. The discovery of the catalytic properties of zeolites occurred in the 1950s, whereas the introduction of zeolites into petroleum refining came in the early sixties. Zeolites revolutionized the important catalytic cracking field and totally penetrated this market in the United States in about one year. In addition, it has been estimated that the introduction of zeolites into catalytic cracking produces an annual savings of over 200,000,000 barrels of oil as compared to quantities that would be required if the older amorphous silica-alumina catalysts were still being used in cracking (290). A number of books and articles have surveyed zeolite catalysis and provide a comprehensive summary of the subject (291–297).

Only a small percentage of all the possible zeolite structures has been discovered (298,299). A probable deterrent to progress in new structure discoveries is the researchers' incomplete understanding of crystallization mechanisms. It certainly is not the case that no advance in crystallization theory or in the understanding of mechanisms has occurred in the last 30 years; indeed, a number of studies have stressed mechanistic and structural information (300). Mechanisms of zeolite structural formation and assignment of the fundamental infrared vibrational modes have been investigated for over 25 years. Perhaps most of the studies have been performed on the synthesis of zeolite A and faujasite. Such a concentration of effort is partly due to the ease of synthesizing these zeolites in highly pure and crystalline forms. Raman spectroscopic measurements have also concentrated on these two zeolites, with SiO_2/Al_2O_3 ratios typically in the range of 1 to 4. This section will concentrate on Raman characterization of zeolites; it also will briefly discuss recent investigations of zeolites having SiO_2/Al_2O_3 ratios greater than five.

Zeolite Synthesis and Structure

McNicol, Pott, and Lloss (301), in the early 1970s, looked at the characteristic Raman vibrational modes of zeolites during synthesis and studied zeolite synthesis as a function of the adsorption of molecules and the type of zeolite to be synthesized. Their study still is used as a reference to discuss two possible mechanisms of zeolite formation that have spectroscopic backing. These mechanisms are: (1) the dissolutuion of an amorphous aluminosilicate gel and formation of a soluble, zeolite precursor aluminosilicate species (302–304); and (2) the polymerization–depolymerization of a solid gel with the subsequent formation of seed nuclei and crystals (305). For example, the Raman results quoted by McNicol did show significant alteration in the solid phase, whereas no changes were observed in the liquid phase. Throughout the induction and growth period for the crystallization of zeolite A, the Raman signal of the N-C symmetric stretching mode of $(CH_3)_4N^+$ in solution was reported to be unaltered in shape, intensity, and position. Upon crystallization of the zeolite, the N-C stretching mode appeared approximately 15 cm^{-1} higher, close the the N-C stretching frequency of $(CH_3)_4N^+$ in sodalite. Hence, the incorporation of $(CH_3)_4N^+$ in the sodalite cage perturbed the N-C vibrational mode; after observation of this perturbed vibration, crystallization of the zeolite was faster, implying an "autocatalysis of crystal growth" (301). Unfortunately, no Raman spectra were published in this important paper; they would have been very informative for spectroscopists performing zeolite synthesis research at the present time.

Prior to McNicol's study, Aiello, Barrer, and Kerr (306) had described the presence of amorphous "lamninae" precursors to zeolite crystallization. Shortly thereafter, Angell showed that quality Raman results could be obtained on zeolites with SiO_2/Al_2O_3 ratios as large as ten (307). Angell and Schaffer previously had investigated zeolites with infrared spectroscopy (308); so Angell's Raman studies were a logical and important extension of molecular vibrational investigations on zeolites.

Before Angell published his study, only one paper, by Hendra, Horder, and Loader (309), had been published that used Raman spectroscopy to monitor the adsorption of gases by zeolites. These Raman investigators noted the difficulty of obtaining spectra having low background. Indeed, two other studies had also commented on the difficulty of obtaining good-quality Raman vibrational spectra while using metal oxide or silaceous samples (310,311). A technique that can decrease such background in these studies has come to be known as "burnout" of samples. This "burnout" is simply the exposure of the sample under investigation to high incident laser irradiation for periods of seconds to several minutes. Even now, over 15 years since these early Raman papers on zeolites, the origin of the excessive fluorescent background is not understood. The size of particles in the zeolite, specific OH surface groups, carbon or hydrocarbon impurities, and trace impurities have all been suggested as causes of such fluorescence. In addition, the type of treatment used on samples to eliminate excessive background has been found to differ, depending on the nature of the sample. For example, it had been found that high temperature treatment under oxygen atmospheres would eliminate excessive background in silica and silica-aluminas; but such treatment does not always decrease background during Raman data acquisition on zeolites (310–312). Perhaps the superior adsorption capacities of zeolites or their microporosity plays a role in the inability to acquire good-quality spectra. Most Raman studies have been performed with the zeolite exposed to atmospheric conditions during data acquisition. It may be beneficial to perform Raman data acquisition to zeolites while they are under controlled environments.

Zeolite A consists of truncated octahedra connected by oxygen atoms between four-membered rings of the octahedra. Stacking these octahedra in a close-packed fashion with alternating eight- and four-member ring connections forms a repetitive crystalline unit known as sodalite. A naturally occurring mineral, sodalite also can be synthesized in highly pure form to produce catalytic and phosphorescent materials (313,314). Angell first reported the Raman spectra of sodalite and related a number of its Raman bands to Si-O bending and stretching modes in quartz (307). The position of the most intense band in quartz, near 465 cm^{-1} and due to Si-O bending (315), is very close to the position of the strongest band in sodalite, chabazite, and offretite (307,314). The position of this Raman band in zeolites has not been found to depend on the SiO_2/Al_2O_3 ratio even though infrared spectra of the same zeolites show shifting as a function of the SiO_2/Al_2O_3 ratio (316). However, the shape and multiplicity of the ca. 460 cm^{-1} band are in agreement with what can be expected from a group theory interpretation. For example, zeolites A, X, and Y and sodalite contain unit cells of cubic symmetry and correspondingly have a strong and narrow Si-O bending mode. In chabozite, which has a trigonal unit cell, the ca. 460 cm^{-1} band is broad and consists of at least three bands. In general, these early studies could not differentiate between distinct Si-O and Al-O modes, primarily because of a lack of exact group theoretical calculations and the small data base of Raman spectra from which comparison could be made. In crystalline alumina oxide the strongest Raman bands were found to be near 380 and 640 cm^{-1}. Although this information was of some value in deciphering zeolite spectra, it was not sufficient to provide unambiguous assignments.

Very few Raman results on zeolites and zeolite synthesis were published between the early seventies and the early eighties. Many zeolite researchers tend to rely on the interpretation of Barrer et al. (302,306) that the polymerization of possibly complex ions was due to the dissolution of the gel. Recent Raman results have also suggested the presence of solute aluminosilicate species (317,318). This Raman information is in agreement with Si NMR data (319,320). However, as discussed by Angell and Flank (321), Raman data also have shown changes

Table 4.11. Summary of Raman vibrational bands from the solid and liquid phases during synthesis of zeolites A, X, and Y. (Ref. 322, By permission of the publishers, Butterworth & Co.) (Publishers Ltd.)

ZEOLITE	TIME	LIQUID BAND (CM^{-1})	SPECIES ASSIGNMENT
A	0	621	$Al(OH)_4^-$
	1	623	$Al(OH)_4^-$
	2	621	$Al(OH)_4^-$
	3	—	—
X	0	621	$Al(OH)_4^-$
		777	$SiO_2(OH)_2^{2-}$
	2	448, 936	$SiO_2(OH)_2^{2-}$
		600	Disilicate
		777	$SiO_2(OH)_2^{2-}$
	8	448, 936	$SiO_2(OH)_2^{2-}$
		601	Disilicate
Y	0	620	$Al(OH)_4^-$
	3	777	$SiO_2(OH)_2^{2-}$
		598	Disilicate
	12	777	$SiO_2(OH)_2^{2-}$
		600	Disilicate
		SOLID	
A	0	—	Amorphous
	2	494	Zeolite A
	3	281, 347	
		405, 415	Zeolite A
		494	
X	0	—	Amorphous
	2	—	Amorphous
	8	291, 376	Zeolite X
		512	
Y	0	—	Amorphous
	3	—	Amorphous
	12	298, 369	Zeolite Y
		512	

in the liquid during zeolite synthesis, thus providing evidence for solution transport mechanisms. As a result, Roozeboom, Robson, and Chan (322) recently reexamined the synthesis of zeolites A, X, and Y with the use of Raman spectroscopy. Positions and relative intensities of the major peaks that were observed are shown in Table 4.11. For zeolite A, the positions of the Raman bands of the solid phase were in general agreement with those presented by Angell and Flank (321). The fact that a significant shift in these positions was observed during crystallization suggested that the crystals grew from solid precursors. In addition, spectra of the liquid phase contained only one peak, at 621 cm^{-1}, which disappeared at approximately three hours after initiation of synthesis; the position of this peak was in agreement with that for an $Al(OH)_4^-$ anion (323). Hence, it was conjectured that $Al(OH)_4^-$ in the liquid phase was being consumed during synthesis. The same effect was observed to occur in the Raman spectra of the liquid phase during synthesis of zeolites X and Y, whereas the spectra of the solid phase of X and Y were similar to each other but different from the spectrum of zeolite A. All of the Raman spectra of the solid phases during synthesis of these zeolites contained a broad background band structure during the first four hours of synthesis; such spectral behavior commonly is found in amorphous aluminas and silicas and may be related to a large variation in Si-O and Al-O bond lengths and angles in the yet poorly crystallized solid phase.

The discussion of Raman data by Roozeboom et al. (322) used the recently published data of Guth et al. (317) on the Raman spectra of various silicate ions, including $SiO_2(OH)_2^{2-}$ monomeric, dimeric, and cyclic-trimeric anions. With these data, not available for previously reported studies on the crystallization of A, X, and Y zeolites, Roozeboom et al. (322) concluded that solution transport was dominant during zeolite synthesis. This Raman data acquisition was hampered by an intense fluorescence background.

Dutta and coworkers (324–327) recently published a series of papers on the Raman spectra of zeolites A, X, and Y. In their study of the influence of cation types on the molecular vibrations of zeolite A (325), it was found that of the cations Na, K, Tl, and Li, only Li perturbed the zeolite structure to a degree observable by use of Raman spectroscopy (Figure 4.25). This study was one of the first to use group theoretical calculations on zeolite A for discerning the symmetry, number, and activity of the vibrational modes. In addition, previously published infrared data on zeolites and preliminary Raman data of de Kanter and coworkers (328) were used to make band assignments to specific aluminosilicate species in the zeolite. Using the space group $O_6/_h$ and the correlation method described by Fateley et al. (329), the vibrational modes were classified as the following:

$$11A_{1g} + 11A_{2g} + 22E_g + 25F_{2g} + 7A_{2u} + 7A_{2u} + 28F_{2u} + 29F_{2u}$$

With further simplification, this vibrational mode description could be broken down to a total of 30 Raman active modes including: $6A_{1g} + 12E_g + 12F_{2g}$. However, only 9 of these Raman modes were observed.

To enhance their group theoretical study, the spectra of sodalite (307,313,330), which is the building block of zeolite A, and the spectra of silica (331), calcium aluminate, anorthite ($CaAl_2Si_2O_8$) (332), and other naturally occurring minerals

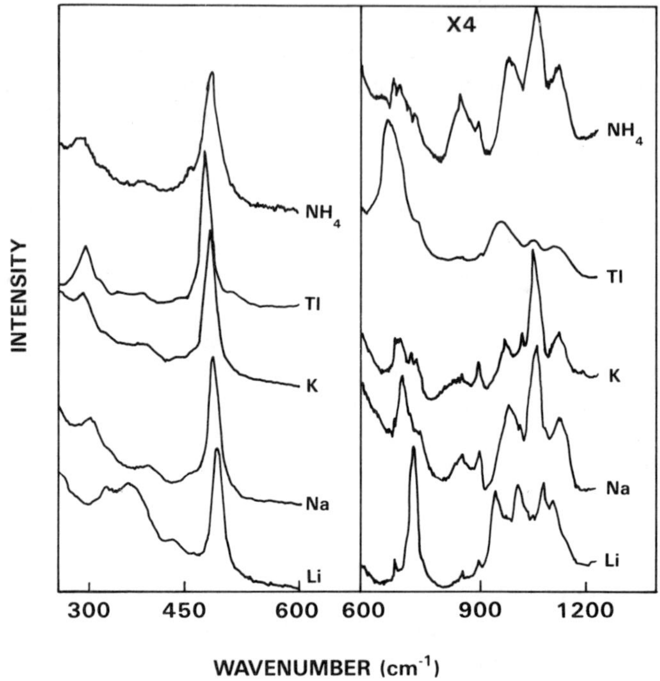

Figure 4.25. Raman spectra of hydrated zeolite A exchanged with Li^+, Na^+, K^+, Tl^+, and NH_4^+ ions, laser line was 457.9 nm, laser power 150 mW, slit width 5 cm^{-1} (324). (Reprinted with permission from *The Journal of Physical Chemistry*. Copyright 1985 American Chemical Society.)

(333) were scrutinized. Such comparative data analyses suggested that Raman bands of zeolite A in the 800 to 1000 cm^{-1} region were the result of Si-O stretching modes of the SiO_4 tetrahedra, whereas the 600 to 800 cm^{-1} region contained primarily Al-O vibrational modes and mixed Si-O and Al-O modes. Below 600 cm^{-1}, the vibrational modes were described as the result of SiO_4 bending and Al-O bending modes. The weak intensity within the latter modes, near 335 and 405 cm^{-1}, was related to residual ionicity of the Al-O bonds.

The study of Dutta and Barco (325) was one of the first to look systematically at the perturbational effects of cations on the Raman vibrational modes of zeolites. It was determined that Li, having a relatively small ionic radius and a correspondingly large coulombic field, caused a relatively large shift in the Si-O and Al-O vibrational positions. Such significant shifts were not generally observed as a function of the concentration of K, Na, or Tl. The fact that no linear change in vibrational frequencies as a function of cation size was observed suggested that a singular electrostatic interpretation of the interaction of the cation with

Si-O-Al units in the zeolite was possibly an oversimplification. However, it was argued that the coulombic field of Li pulled the zeolitic oxygens toward the cation, thereby distorting the Si-O, Al-O, and Si-O-Al bond angles and energies. As a result, and in agreement with previous infrared studies on sodalite (334), the Raman Si-O stretching mode frequencies decreased in frequency. In addition, the effects of NH_4^+ cations on the spectra of zeolite A were outlined briefly. Here, even though the ionic size of NH_4^+ is only 60% of that for Na, no significant differences were found between spectra of NH_4- and Na-A. Some relative shifts in frequency of the bending modes were explained in terms of the effects of H-bonding on T-O bonding (335).

Structure sensitivity in the Raman bands for zeolite A has been observed (325); the vibrational spectra of silica and aluminosilicate glasses (336,337) were used to suggest an origin of some of its Raman bands. Quantitative X-ray diffraction data were used to define bond lengths and angles within the zeolite as a function of cation substitution and to provide an indication of which Raman bands would be affected by such substitution. The Si-O stretching modes in the 900 to 1200 cm^{-1} region were found to be most sensitive to cation interaction. The disappearance of the 1095, 1034, and 962 cm^{-1} bands as K was substituted for Li, and the corresponding appearance of 1086, 1057, 996, and 937 cm^{-1} bands, were interpreted as a result of Li substitution into S_{II} sites, subsequent distortion of the cubic lattice, and then substitution of K into the S_I sites.

Dutta et al. (327,338) also studied the synthesis of zeolites A and Y by examining Raman spectra of the solid and liquid phases of the gel as a function of crystallization time. These results can be compared to the results of Roozeboom et al. (322); however, Roozeboom et al. used a higher temperature during synthesis, thereby increasing the rate of crystallization. Band positions and band assignments below 600 cm^{-1} specific to solid and/or liquid species are similar in these studies (322,338,339). In addition, ^{27}Al NMR was used by Dutta et al. to confirm the structure of the aluminosilicate species. These NMR and Raman data, and correlations between the Raman frequencies in the gel with those in the naturally occurring mineral having known ring size and structure (340), provided rather solid arguments in favor of a depolymerized $Al(OH)_4$ ring structure in the gel. This cation was considered to be responsible for the crystallization process.

In the case of zeolite Y synthesis (327), Raman and NMR data were compared to the same type of data from well-characterized and naturally occurring minerals (333,341,342). By investigating liquid and solid phases of the gel during crystallization, it was shown that the solid amorphous phase consisted of six-membered aluminosilicate rings—the building block for formation of zeolite Y. Contrary to zeolite A synthesis (338), the Raman data for zeolite Y suggested a need to have polymeric-condensed silicate units for crystallization to occur.

The effect of zeolite structure and cage size on the vibrational frequencies of

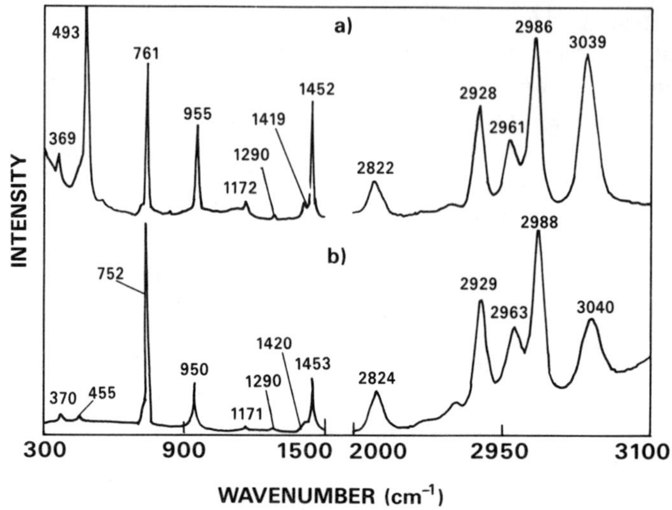

Figure 4.26. Raman spectra of the tetramethylammonium cation (a) as trapped in the cages of sodalite, and (b) in an aqueous solution (1 *M*). Laser excitation was 457.9 nm, power at sample 30 to 50 mW, and slit width 6 cm^{-1} (326).

tetramethylammonium (TMA) template cations was investigated by Dutta, Barico, and Shieh (326). The data for TMA/sodalite are shown in Figure 4.26 and for TMA/ZK-4 in Figure 4.27. Corresponding assignments of bands in these spectra are listed in Table 4.12. The spectra of ZK-4, a high-silica analog of zeolite A, show an increase in intensity of a band at 755 cm^{-1} relative to the intensity of a band at 767 cm^{-1} as the Si/Al ratio increases. These two bands are a result of C-N symmetric stretching vibrations of TMA in the supercages of the zeolite. Changes in band positions and cage effects on the C-N symmetric stretching mode on sodalite were explained in terms of the tightness of fit of TMA in the cages. Such a close fit produces different degrees of perturbation on the various vibrational modes, where, for example, a symmetric stretching vibration would require more space for unpertured motion than asymmetric or bending modes. Hence, Raman spectroscopy was said to be able to define specific zeolite structural effects on the templates. These conclusions are in agreement with high-resolution NMR data (343).

Effects of the ZSM-5 cage on its synthesis template, tetrapropylammonium (TPA), were examined by Baranska, Czerwinska, and Labudzinska (344). The CH_3 and CH_2 stretching vibrational frequencies were lower by about 19 cm^{-1} for TPA in the zeolite as compared to the ion in solution. This decrease was similar to that discussed by Dutta et al. (326) for TMA in soldalite, ZK-4, and zeolite A. Additionally, the intensity ratio, $I(CH_3)/I(CH_2)$, for the symmetric

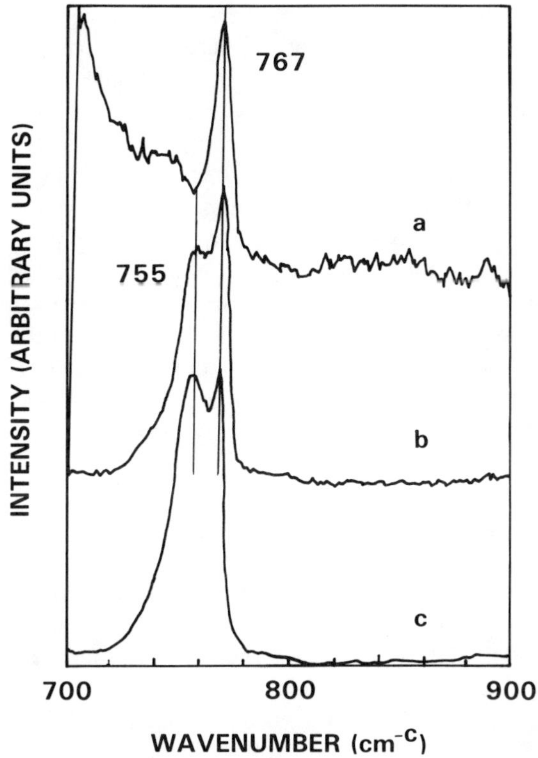

Figure 4.27. Raman spectra of zeolite ZK-4 with differing Si/Al ratios: (a) 1.08, (b) 2.49, and (c) 2.63 (326).

Table 4.12. Assignments of vibrational bands of TMA in sodalite and aqueous solution (326).

ASSIGNMENT	TMA(SOLUTION) (CM^{-1})	TMA(SODALITE) (CM^{-1})
T_2 : CH_2 asym. stretch	3040	3039
A_1 : CH_3 sym. stretch	2988	2986
T_2 : CH_3 sym. stretch	2929	2928
E : CH_3 asym. deformation	1453	1452
T_2 : C—N asym. stretch	950	955
A_1 : C—N sym. stretch	752	761
T_2 : C—N deformation	455	—
E : C—N deformation	370	369

stretching modes was 0.8 in the zeolite versus 1.3 in aqueous solution; deformation modes of CH_3 and CH_2 were not perturbed. Perhaps, as suggested by Dutta et al. the size of the TPA is such that vibrational modes associated with symmetric stretching interact more fully with the surrounding zeolite structure than other vibrational modes that do not involve a symmetric enlargement of the size of the trapped template. The data of Baranska et al. (344) and the Raman and infrared data detailing the internal modes of ZSM-5 and silicalite (345) provide a relatively sound basis for describing ZSM-5 synthesis and molecular structure. In the latter report, data acquisition at temperatures as low as 15°K allowed resolution of vibrational bands that were not resolvable at room temperature. Such low temperature information, especially in the case where the crystal structure does not change at lower temperatures, is relatively easy to obtain and interpret. Low temperature data would also enhance the application of group theory and identification of the origin of the fundamental modes of other zeolites.

Zeolite Deactivation

Raman spectroscopy generally has not been used to delineate deactivation mechanisms of catalysts, a fact that is not surprising in the sense that deactivation associated with catalytic poisons (e.g., carbon or sulfur) normally produces black materials that absorb the incident laser radiation. By sensitive adjustments of experimental equipment it may be possible to use such black materials under Raman conditions. Indeed, as described below, the use of Raman spectroscopy to provide structural information about carbon as it is deposited may be an attractive possibility. The main part of this discussion, however, will concentrate on the implications of recent Raman studies on metal poisoning of FCC catalysts.

V and Ni Poisoning

The catalyst used in the largest volume and having the highest value per year in petroleum refining is the FCC catalyst. Its primary component is generally a rare-earth, ion-exchanged zeolite X or Y, or dealuminated Y, that is mixed or composited with other needed structural and chemical modifiers such as clays (kaolin) and amorphous silica-alumina (AAA). Metals such as Ni, V, and Cu in the feed oil stream passing over the FCC cause irreversible deactivation of the FCC even at relatively low concentrations in the feed (346). The mechanism of deactivation can be associated with (1) destruction of the zeolite crystallinity, (2) physical blockage of active sites through pore size restrictions, and (3) chemical poisoning of active sites through active site–poison association. The results of such poisoning in the refinery are increased production of unwanted light gases and carbon and a subsequent decrease in gasoline make. The trend in oil production is toward increased levels of such metal poisons, which have

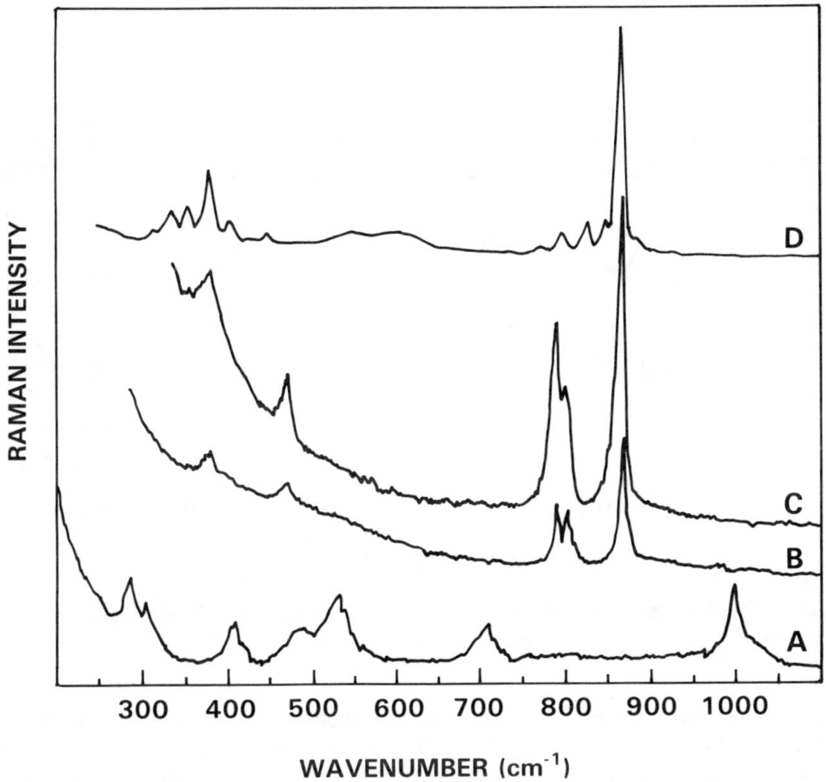

Figure 4.28. Raman spectra of 5% V on (a) HY zeolite and (b) calcined rare-earth exchanged Y zeolite (CREY) after steam aging, with (c) and (d) reference spectra for $CeVO_4$ and $LaVO_4$, respectively (349).

to be handled satisfactorily in the refinery. The effects of metals on FCC have been examined in some detail (297,346–358). Still, there is no consensus on the mechanisms of the interaction of Ni, V, and Cu with the FCC that causes the destruction of zeolite crystallinity. Recent Raman spectroscopic data, and other complementary data, have offered concrete evidence pertaining to FCC and dual function FCC (DFFCC) catalyst deactivation.

Occelli and Stencel (349) investigated the form of V on HY, CREY, kaolin, and amorphous aluminosilicates after calcination and steaming of these catalysts. Figures 4.28 and 4.29 show Raman spectra of these V-contaminated catalysts and the corresponding spectra of $CeVO_4$, $LaVO_4$, and V_2O_5. On steam-aged kaolin, amorphous aluminosilicate, and lanthinide-free HY, vanadium is present mainly as V_2O_5. In contrast, the Raman spectrum of V-contaminated CREY exhibits bands at 869, 803, 766, 470, and 370 cm^{-1} that are in excellent agreement

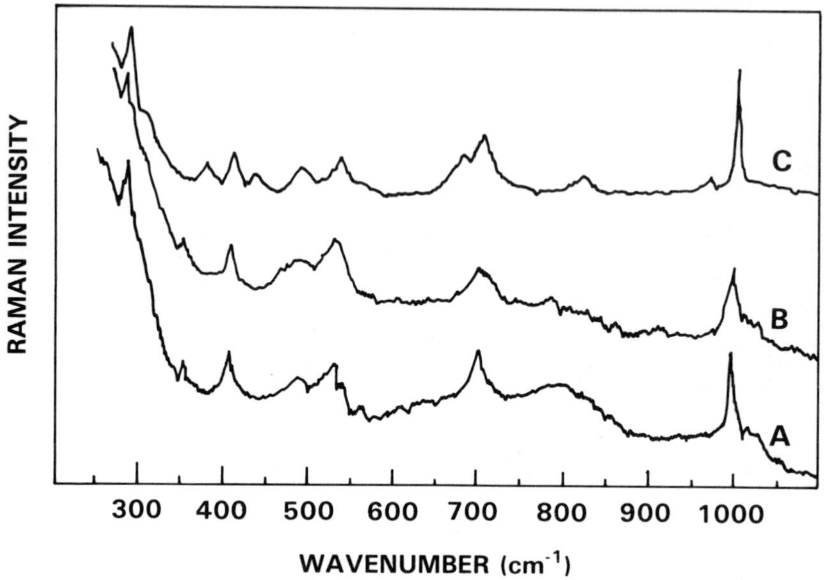

Figure 4.29. Raman spectra of 5% V loaded on (a) kaolin, (b) gel (AAA alumina) after steam aging, and (c) reference spectra for V_2O_5 (349).

with the spectrum of $CeVO_4$. Although the concentrations of the lanthanides Ce and La in the CREY were 7.6% as Ce_2O_3 and 4.0% as La_2O_3, respectively, $LaVO_4$ formation could not be observed. The identification of $CeVO_4$ has been confirmed with X-ray diffraction.

To look at the dependence of $CeVO_4$ formation on the V-loadings, the steamed-aged and V-contaminated CREY samples were homogeneously mixed with known quantities of $NaNO_3$. Raman data were obtained on these mixtures, in which the intensity due to the NO_3 symmetric stretching vibrational band at 1357 cm^{-1} was compared to the VO_4 symmetric stretching vibration at 875 cm^{-1}. The resulting linear increase in $CeVO_4$ as a function of V-loading is shown in Figure 4.30. The lower limit of $CeVO_4$ detection is at approximately 0.7% V. This lower limit for detecting $CeVO_4$ was in close agreement with results of a concomitant X-ray diffraction study (349). Although Raman spectroscopy of catalysts is not generally thought quantitative, studies have proved that specially designed and relatively simple experimentation can provide quantitative data acquisition (46,71).

The linear dependence of $CeVO_4$ formation on V-loading at concentrations above approximately 0.7% V implies that Ce–V interactions continue to occur at loadings above which X-ray diffraction indicates a complete collapse of the zeolite structure. Below 0.7% V the sensitivity of Raman spectroscopy was

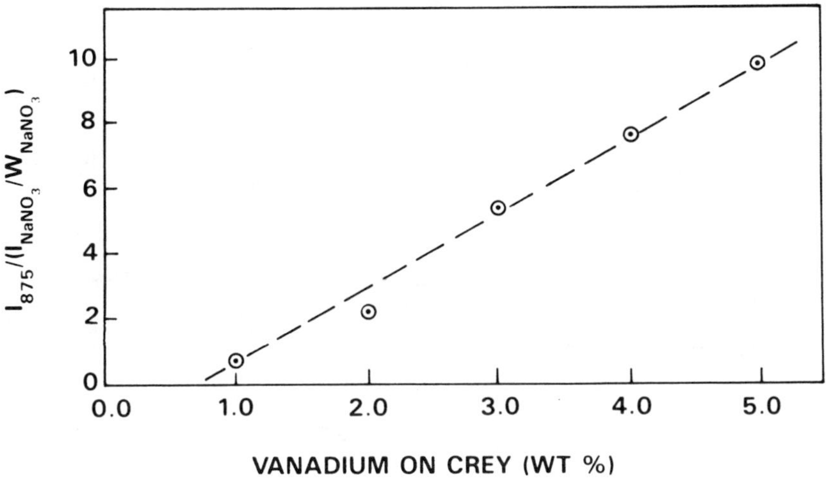

Figure 4.30. Quantitative plot of $CeVO_4$ formation as a function of V-loading (349).

somewhat inadequate to identify $CeVO_4$. However, because extraction of Ce from the CREY would be expected to lead to a decrease in zeolite crystallinity, $CeVO_4$ also should be formed at loadings of less than 0.7% V. The fact that no $LaVO_4$ was formed suggested that either the thermodynamics are not favorable, or sites of incorporation of Ce in the zeolite Y are more accessible to V attack than are the sites of La incorporation. The stoichiometry of Ce:V at a 5% V loading is near 1:1, implying that the characterization of CREY with greater than 5% V may show the appearance of $LaVO_4$ bands. In the case of LaY zeolite, $LaVO_4$ was formed upon steaming of the catalyst.

Resistance to metal poisoning in an FCC can be achieved by adding to the feed metals Sb (350), Bi (351), Sn (352), Sn–Sb combinations (353), Ba (354), P and B (355). The addition of these compounds is believed to cause the formation of inert compounds on the surface of the FCC. Recent approaches to scavenging of poisons such as V or Ni have relied on the addition of certain diluents to the FCC composition (356–359). These diluents, which are inorganic oxides, are not so environmentally objectionable as are certain other compounds (350–355). Raman spectroscopy has been important in defining the interaction of V with these dual function fluid catalytic cracking (DFFCC) catalysts.

One diluent that has been used to scavenge V, and thereby decrease it deleterious effect on the zeolite, is sepiolite [$Mg_8Si_{12}O_{30}(OH)_4$]. Sepiolite is a fibrous clay mineral composed of tetrahedral chains of silica lying in planes, joined together (through shared oxygens) by magnesium (and aluminum) atoms in octahedral coordination (360). The Raman spectra of 5% V/sepiolite after exposure to steam aging conditions are similar to those encountered in the steam

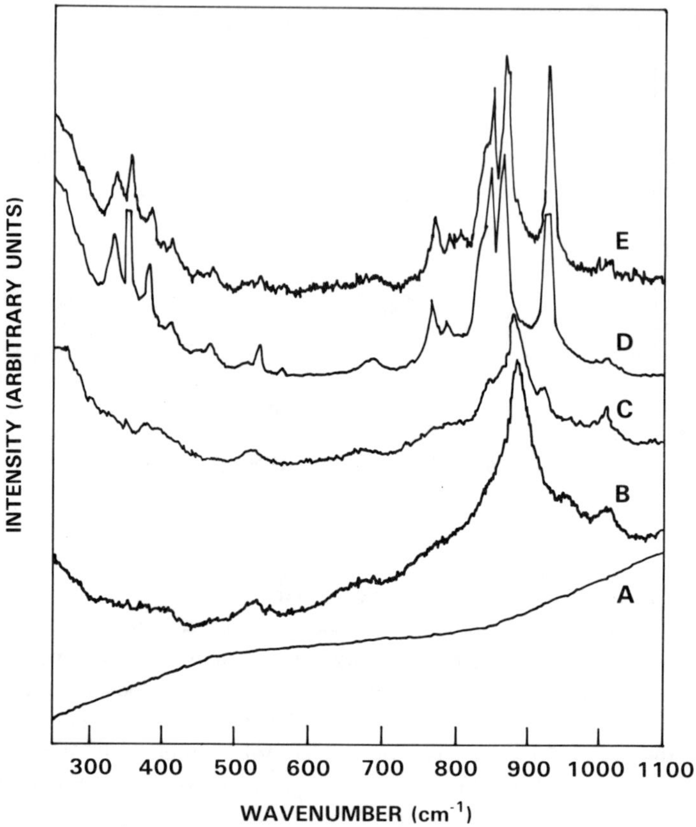

Figure 4.31. Raman spectra of: (a) Spanish sepiolite calcined at 800°C/1 h; (b) Spanish sepiolite containing 5% V, calcined at 800°C/1 h; (c) sample (b) steamed; (d) IMV sepiolite containing 5% V, calcined at 800°C/1 h; and (e) sample (d) steamed (361).

stripper region of a catalytic cracking reactor; these spectra are shown in Figure 4.31 (361,362). Band positions and intensities, together with Raman data for reference compounds, are listed in Table 4.13, whereas the corresponding spectra of the reference compounds are shown in Figures 4.32 and 4.33. By examination of these spectra it is obvious that V/sepiolite does not contain V_2O_5 or tetrahedral VO_4 species. The positions of bands at 885, 845, and 525 cm^{-1}, present in V_2O_7-containing samples such as CdV_2O_7 (363), are in general agreement with those bands in the spectra of the V-containing sepiolite. Furthermore, two bands, one at 959 cm^{-1} and the other at 885 cm^{-1}, in the spectra of CaV_2O_6 and MgV_2O_6 are also in close proximity to the band positions observed in the V-contaminated sepiolite.

Table 4.13. Raman spectra of V-loaded (5%) sepiolites and reference compounds.

SAMPLE	CM^{-1} (INTENSITY)
V_2O_5	997 (10), 846 (6), 788 (4.5), 729 (7.5), 670 (6), 605 (8)
$[VO_4)^{3-}]$*	827 (10), 780 (2)
$[Pb_5(VO_4)_3Cl]$*	829 (10), 799 (2)
MgV_2O_6	921 (10), 836 (3), 730 (4), 523 (0.5)
CaV_2O_6	959 (10), 885 (5.5), 723 (1.5), 555 (1.5)
$[Cd_2V_2O_7]$	877 (10), 848 (3), 820 (2), 789 (0.5)
$[V_2O_7)^{4-}]$*	877 (10), 850 (2), 810 (2), 503 (2)
α-$Mg_2V_2O_7$	948 (5.5), 919 (6.5), 902 (10), 873 (3.5), 842 (3), 723 (1.5), 631 (1), 621 (1), 570 (0.5)
β-$Mg_2V_2O_7$	1014 (3), 920 (1), 898 (5), 882 (10), 860 (3), 845 (4.5), 788 (1.5), 740 (1), 523 (1.5)
MgV_2O_6–$Mg_2V_2O_7$	1014 (2), 923 (10), 882 (6.5), 845 (4), 840 (4), 732 (4), 522 (1)
$Mg_3V_2O_8$	863 (10), 827 (2)
5% V/Sp-Sep, steam	1014 (1.5), 952 (2.5), 885 (10), 850 (sh), 525 (1)
5% V/Sp-Sep CA	1014 (2.5), 925 (4.5), 898 (sh), 885 (10), 850 (5), 525 (1)
5% V/IMV-Sep, steam	1013 (1), 929 (9.5), 868 (10), 848 (9), 835 (5), 773 (3)
5% V/IMV-Sep CA	1015 (0.5), 929 (10), 867 (9), 849 (8.5), 836 (5.5), 788 (2), 770 (3), 690 (0.5), 530 (1)

sh = shoulder, CA = calcined.
*References, 83, 84.

The form of the V compound on sepiolite is dictated by the purity of the sepiolite. For example, it was found (361) that the V compound forming on pure sepiolite (designated Spanish sepiolite in Figure 4.31) was mainly a MgV_2O_6 + beta-$Mg_2V_2O_7$ mixture, whereas in an impure sepiolite (containing $CaCO_3$ and $MgCO_3$ impurities) the vanadium also formed $Mg_3V_2O_8$. The variations in the intensity of the Raman bands for the various V/sepiolite samples in Figure 4.31 were ascribed to variations in the concentrations of the Mg_1-, Mg_2-, and Mg_3-vanadates. Although the data in Figures 4.31 through 4.33 are for the V/sepiolite and not for a V/Y zeolite with sepiolite, the migration of V under catalytic cracking conditions is well known (362). Hence, the interaction of V from the feed stream with the DFFCC would entail scavenging of the V from the feed by the sepiolite before it destroys zeolite crystallinity.

The mechanism of V attack on the zeolite component of an FCC also can be discussed with respect to the data shown in Figures 4.31 through 4.33. For example, Wormsbecher, Peters and Maselli (348) have proposed that at the condition used to regenerate cracking catalysts (700°C, 20% steam), a volatile H_3VO_4 species is formed, causing the destruction of the zeolite. Griffith and coworkers (363,364), in studying the Raman and infrared spectra of aqueous vanadates, were able to demonstrate the presence of $(VO_3)_n$ at pH less than 9.5, $(HV_2O_7)^{3-}$ with the pH in the range 9.5 to 11.9, and $(V_2O_7)^{4-}$ and $(HVO_4)^{2-}$ at

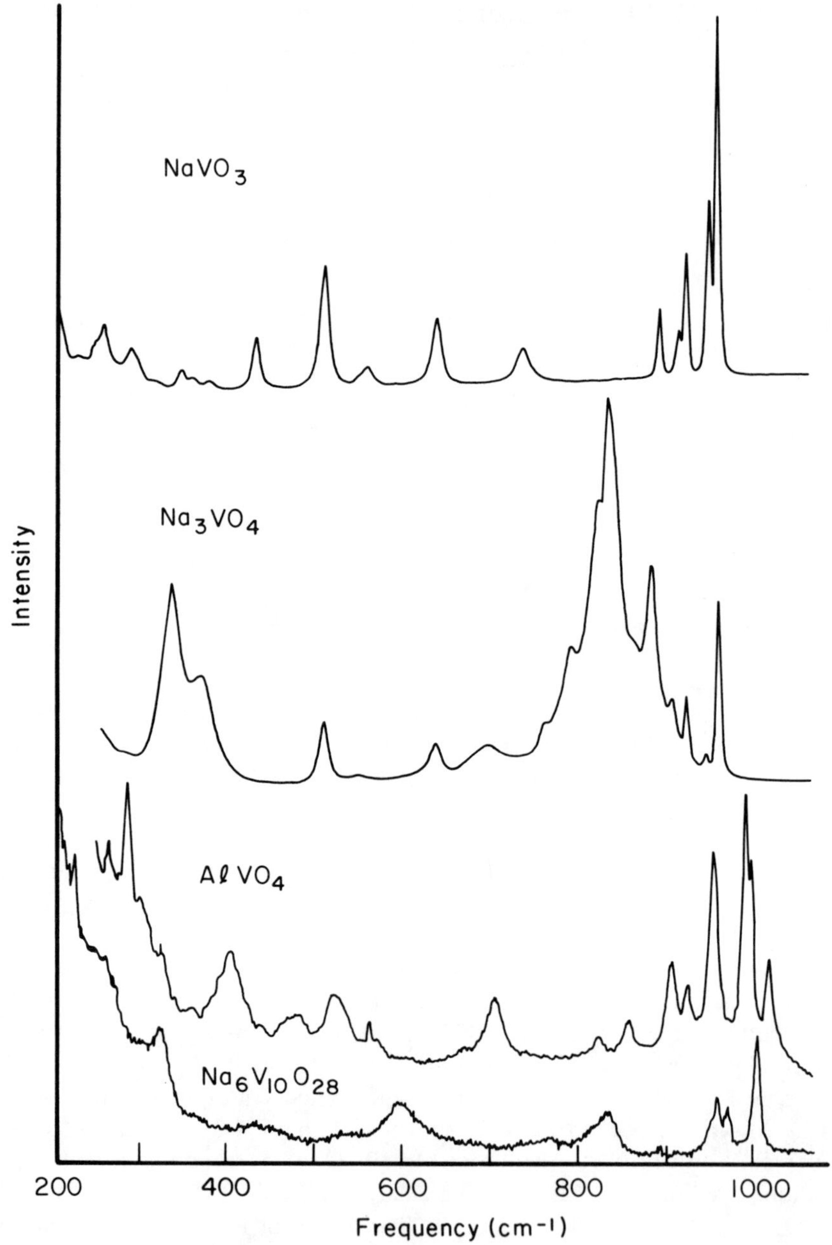

Figure 4.32. Raman spectra of reference compounds used in analyzing V-contaminated catalysts.

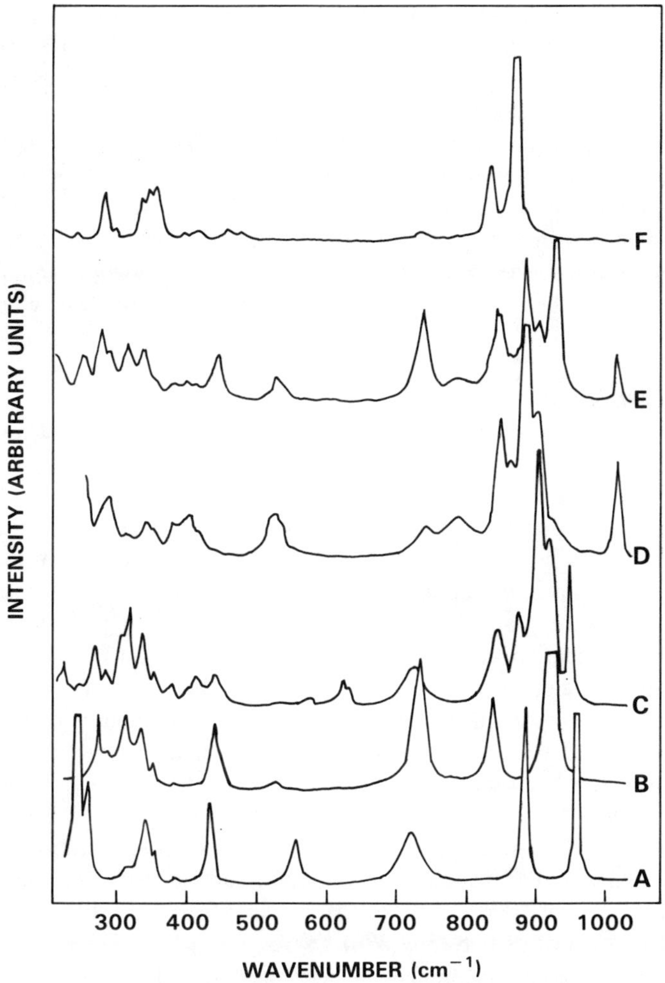

Figure 4.33. Raman spectra of unsupported: (a) CaV_2O_6; (b) MgV_2O_6; (c) alpha-MgV_2O_7; (d) beta-$Mg_2V_2O_7$; (e) a MgV_2O_6-$Mg_2V_2O_7$ mixture; (f) $Mg_3V_2O_8$.

a pH of 11.9 to 14. At a pH greater than 14 (using 10 *M* KOH solutions), the VO_4^{3-} ion was observed; its spectrum was characterized by a sharp Raman line at 827 cm^{-1} and a weak line at 340 cm^{-1}. Evidence of vanadic acid (H_3VO_4) has not been detected in any of the sepiolite samples. It may be doubtful that the strong alkaline conditions required to form the VO_4^{3-} ions could be achieved on an FCC surface. Additionally, the Raman bands of the DFCC in Figure 4.32 at 1014 cm^{-1} and 952 cm^{-1} are in close agreement with those found in the mineral hummerite [K_2Mg_2 $(V_{10}O_{28})\cdot 16H_2\cdot O$]. Hence, considering only Raman data of the sepiolite samples, Raman spectra of standard compounds, and Raman data on the effect of anion distortion and H-bonding on V-O stretching frequencies (363–366), it is suggested that the mechanism of interaction of V with the zeolite component in an FCC involves protonated species such as $(HV_2O_7)^{3-}$ or $(HV_{10}O_{28})^{5-}$.

Besides the interaction of V with sepiolite, Raman spectroscopy was used to study the interaction of Ni with sepiolite (361). These results suggested that the interaction of Ni with Mg-silicate lattice of sepiolite occurs through the formation of $NiSiO_3$. A complementary X-ray photoelectron spectroscopic and X-ray diffraction study of the same samples produced the same conclusions as did the Raman study (361).

Carbon Deposition

The deposition of carbon on catalysts can be a significant factor in their deactivation. It may be possible by the use of Raman spectroscopy to elucidate specific forms of the carbon and their average particle size. For example, the characterization of graphitic and nongraphitic carbon by Raman spectroscopy was recently investigated by Johnson (367). The main Raman peak from graphite was the E_{2g} interplanar vibrational mode at ca. 1580 cm^{-1} (368). The introduction of disorder into the graphitic structure produced broadening of the 1580 cm^{-1} band with the creation of bands at approximately 1600 cm^{-1} and 1360 cm^{-1}. Using well-characterized carbons, Johnson was able to define a relationship between an average size of the basal dimension (L_a) in graphitic carbon and the intensity ratio of the 1360 and 1580 to 1600 cm^{-1} bands, shown in Figure 4.34. Such information and other data on the intensity ratios of these Raman bands (368–370) show that very small carbon particle sizes ($< \sim 2.0$ nm) can be monitored by Raman spectroscopy. The possibility of characterizing such small carbon particles implies that characterization of carbonaceous deposits is readily achievable, even for "particle" sizes that are less than the approximate 7.0 nm optical skin depth of carbon (369,370).

Graphites and carbon materials have been investigated by use of Raman spectroscopy by several scientists (368–373). Besides the bands near 1360 and 1580 cm^{-1}, weaker bands are located at 2700 and 2735 cm^{-1} for highly ordered carbon. Upon disordering of the regular hexagonal graphite structure, the 1580 cm^{-1} band also shifts to higher frequencies (374). This behavior has been noted in SERS data of carbon deposited on Au and Ag electrodes/surfaces during pyridine adsorption (375,376). In fact, determining the effect of carbon deposition on the anomalous enhancements observed in SERS has been a major research effort. From SERS and other normal Raman spectroscopic experiments it can be stated that small structural changes in carbons are easily monitored by the use of Raman spectroscopy (377,378).

For deposition of a thin layer ($\leqslant$ 10 nm) of Ag on top of a carbon layer

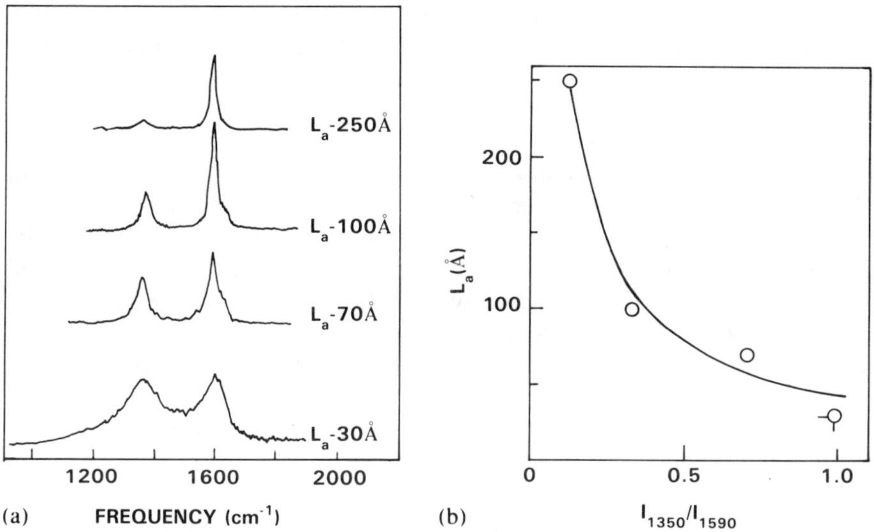

Figure 4.34. (a) Experimental first-order Raman spectra of carbon materials having particles sizes as listed. (b) Plot of the particle size of carbon versus the intensity ratio I_{1350}/I_{1580} (367). (Reprinted from *Microbeam Analysis—1986*, San Francisco Press, 1986, p. 28, by permission).

produced an approximate 100-fold enhancement in the Raman intensities of the carbon peaks (379). This intensity enhancement was more predominant for the ca. 1360 cm^{-1} band than for the ca 1580 cm^{-1} band. Hence, a thin Ag coating produced a more surface-specific Raman spectrum, estimated to originate at a scattering depth of no more than 2.0 nm. Simultaneous with such increased surface sensitivity was the detection of new bands at 1550 and 1140 cm^{-1}. These peaks, if correctly assigned to the C═C and C—C stretching vibrations, respectively, of a surface polyene-like structure (379–382), imply that Raman spectroscopy can be made to be as surface-sensitive for carbon characterization as more familiar surface techniques such as XPS, Auger, and ISS.

The Raman spectroscopic results of Brown and coworkers (1–3) showed substantial carbon deposition on Co-Mo/Al_2O_3 catalysts that had been removed from a coal liquefaction reactor. Unfortunately, there was not much discussion of these carbon data. More recently, Raman spectra of carbon that had been deposited on nickel catalysts used for steam reforming of oil distillates were discussed (383). These spectra showed differences expected for the three forms of carbon—amorphous, vermicular, and platelet (384,385). An analysis of the distribution of the forms of carbon within these catalyst pellets by the use of Raman microprobe instrumentation was also described; the spatial resolution of this instrument was approximately 30,000 nm. Analysis of the size of the carbon particles with respect to the data described by Tunistra and Koenig (368,372) gave results outside the size range for the analyzed particles. However, the form of the carbon deposits was elucidated, and it was found that the rate of carbon deposition depended on the porosity of the catalyst pellets.

The origin of the ca. 1360 cm^{-1} band in carbon deposits is still a topic of research. As noted above, Johnson (367) suggested that its origin was the result of disorder in the graphitic structure. Other possibilities for its origin include amorphous carbon and/or extremely small-sized graphite forms (e.g., pregraphite carbon). This latter possibility has been theoretically studied by Lespade, Jishi and Dresselhaus (386), who have shown that Raman inactive modes are possible for small (~2.5 nm) carbon crystallites because symmetry restrictions no longer necessarily apply. These calculations do show that a Raman mode should appear in the vicinity of the 1360 cm^{-1} band that is observed in some carbon samples. For example, with relaxation of symmetry restrictions, it has been shown that bands at 1620 (A_1), 1580 (A_2), 1530 (A_3), 1450 (A_4), and 1340 cm^{-1} (A_5) should appear (386).

Coke deposition can be a major reason for the deactivation of reforming catalysts. Hence, Espinat et al. (387) undertook a careful study of a series of Pt/Al_2O_3, $Pt\text{-}Ir/Al_2O_3$, $Pt\text{-}Re/Al_2O_3$, and $Pt\text{-}Sn/Al_2O_3$ catalysts. The sensitivity of Raman to monitor coke deposits in the catalysts was found to be excellent, with an ease of obtaining data for coke contents from 0.29 from 27.3%. Raman spectra as a function of the H_2/CH ratio (2.2:8.2) of the coke on the catalyst were presented, with the conclusion that the ca. 1360 cm^{-1} band increased in intensity with increasing H_2/HC ratio. The importance of this H_2/HC ratio with respect to enhanced performance of the bimetallic catalyst as compared to a monometallic reforming catalyst was discussed briefly.

REFERENCES

1. Brown, F. R., L. E. Makovsky, *Appl. Spectrosc. 31,* 44 (1977).
2. Brown, F. R., L. E. Makovsky, and K. H. Rhee, *J. Catal. 50,* 162 (1977).
3. Brown, F. R., L. E. Makovsky, and K. H. Rhee, *J. Catal. 50,* 385 (1977).
4. Villa, P. L., F. Trifiro, and I. Pasquon, *React. Kinet. Catal. Lett. 1,* 341 (1974).
5. Beuchler, E., and J. Turkevich, *J. Phys. Chem. 76,* 2325 (1972).
6. Clark, G. M., and W. P. Doyle, *Spectrochim. Acta 22A,* 1441 (1966).
7. Russell, A. S., and J. J. Stokes, *Ind. Eng. Chem. 38,* 1071 (1946).
8. Richardson, J. T., *Ind. Eng. Chem. Fundam. 3,* 154 (1964).
9. Lipsch, J. M. J. G., and G. C. A. Schuit, *J. Catal. 15,* 174 (1969).
10. Sonnemans, J., and P. Mars, *J. Catal. 31,* 209 (1973).
11. Ashly, J. H., and P. C. H. Mitchell, *J. Chem. Soc. A 1969,* 2730 (1969).
12. Asmolov, G. N., and O. Krylov, *Kintec. Catal. 11,* 847 (1970).
13. Giordano, N., J. C. J. Bart, A. Vaghi, A. Castellan, and G. J. Martinotti, *J. Catal. 36,* 81 (1975).

14. Brown, F. R., R. Tischer, L. E. Makovsky, and K. H. Rhee, Division of Pet. Chem., ACS meeting, Anaheim, CA (Mar. 12–17, 1978).
15. Medema, J., C. van Stam, V. H. J. deBeer, A. J. A. Konings, and D. C. Koningsberger, *J. Catal. 53,* 386 (1978).
16. Knozinger, H., and H. Jeziorowski, *J. Phys. Chem. 82,* 2002 (1978).
17. Beattie, I. R., and T. R. Gilson, *Proc. Roy. Soc. A 307,* 407 (1968).
18. Beattie, I. R., and T. R. Gilson, *J. Chem. Soc. A 1969,* 2322 (1969).
19. Schroder, F. A., B. Krebs, and R. Mattes, *Naturforsch. 276,* 22 (1972).
20. Jeziorowski, H., and H. Knozinger, *J. Phys. Chem. 83,* 1166 (1979).
21. Aveston, J., E. W. Anacker, and J. S. Johnson, *Inorg. Chem. 3,* 735 (1964).
22. Kepert, D. L., *Compr. Inorg. Chem. 4,* 607 (1973).
23. Mattes, R., H. Bierbusse, and J. Fuchs, *Z. Anorg. Allg. Chem. 385,* 230 (1971).
24. Tytko, K. H., and B. Schonfeld, *Z. Naturforsch. 30,* 471 (1975).
25. Fuchs, J., and J. Brudgam, *Z. Naturforsch, 32B,* 853 (1977).
26. Zingg, D. S., L. E. Makovsky, R. E. Tischer, F. R. Brown, and D. M. Hercules, *J. Phys. Chem. 84,* 2898 (1980).
27. Giordano, N., J. C. J. Bart, A. Castellan, and A. Vaghi, *J. Less Common Met. 54,* 387 (1977).
28. Krylov, O, V., L. Y. Margolis, *Kinet. Katal. 11,* 358 (1970).
29. de Beer, V. H. J., and G. C. A. Schuit, in *Preparation of Catalysts,* ed. B. Delmon, P. A.Jacobs, and G. Poncelet, (Elsevier, Amsterdam, 1976), p. 343.
30. Payen, E., J. Barbillat, J. Grimblot, and J. P. Bonnelle, *Spectrosc. Lett. 11,* 997 (1978).
31. Cheng, C. P., and G. L. Schrader, *J. Catal. 60,* 276 (1979).
32. Wang, L., and W. K. Hall, *J. Catal. 66,* 251 (1980).
33. Brunnelle, J. P., *Pure Appl. Chem. 50,* 1211 (1978).
34. Wang, L., and W. K. Hall, *J. Catal. 77,* 232 (1982).
35. Wang, L., and W. K. Hall, *J. Catal. 83,* 242 (1983).
36. Thomas, R., V. H. J. de Beer and J. A. Moulijn, *Bull. Soc. Chim. Belg. 90,* 1349 (1981).
37. Schrader, G. L., and C. P. Cheng, *J. Catal. 80,* 369 (1983).
38. Stencel, J. M., L. E. Makovsky, T. A. Sarkus, J. de Vries, R. Thomas, and J. Moulijn, *J. Catal. 90,* 314 (1984).
39. Leyrer, J., B. Vielhaber, M. I. Zaki, Z. Shuxian, J. Weitkamp, and H. Knozinger, *Mat. Chem. Phys. 13,* 301 (1985).
40. Dixit, L., D. L. Gerrard, and H. J. Bowley, *Appl. Spect. Rev. 22,* 189 (1986).
41. Stampfl, S. R., Y. Chen, J. A. Dumesic, C. Niu, and C. G. Hill, Jr., *J. Catal. 105,* 445 (1987).
42. Luthra, N. P., and W. C. Cheng, *J. Catal. 107,* 154 (1987).
43. Kasztelan, S., J. Grimblot, J. P. Bonnelle, E. Payen, H. Toulhoat, and Y. Jacquin, *Appl. Catal. 7,* 91 (1983).
44. Payen, E., S. Kasztelan, J. Grimblot, and J. P. Bonnelle, *J. Mol. Struct. 143,* 259 (1986).
45. Shimada, H., T. Sato, Y. Yoshimura, J. Hiraishi, and A. Nishijima, *J. Catal. 110,* 275 (1988).
46. Baltrus, J. P., L. E. Makovsky, J. M. Stencel, and D. M. Hercules, *Anal. Chem. 57,* 2500 (1985).
47. Akratopulu, K. Ch., L. Vordonis, and A. Lycourghiotis, *J. Catal. 109,* 41 (1988).
48. Jeziorowski, H., H. Knozinger, P. Grange, and P. Gajardo, *J. Phys. Chem. 84,* 1825 (1980).
49. Stencel, J. M., J. R. Diehl, J. R. D'Este, L. E. Makovsky, L. Rodrigo, K. Marcinkowska, A. Adnot, P. C. Roberge, and S. Kaliaguine, *J. Phys. Chem. 90,* 4739 (1986).
50. Rodrigo, L., K. Marcinkowska, A. Adnot, P. R. Roberge, S. Kaliaguine, J. M. Stencel, L. E. Makovsky, and J. R. Diehl, *J. Phys. Chem. 90,* 2690 (1986).

51. Rocchiccioli-Deltcheff, C., P. Thouvenot, and R. Franck, *Spectrochim. Acta 32A,* 587 (1976).
52. Kasprzak, M. S., S. R. Crouch, and G. E. Leroi, *Appl. Spectrosc. 32,* 537 (1978).
53. Lyham, L., S. J. Cyvin, B. N. Cyvin, and J. Brunvoll, *J. Spectrosc. Lett. 9,* 859 (1976).
54. Marcinkowska, K., L. Rodrigo, P. C. Roberge, and S. Kaliaguine, *J. Mol. Catal. 33,* 189 (1985).
55. Castellan, A., J. C. J. Bart, A. Vaghi, and N. Giordano, *J. Catal. 42,* 162 (1976).
56. Gajardo, P., P. Grange, and B. Delmon, *J. Phys. Chem. 83,* 1771 (1979).
57. Keggin, J. F., *Proc. Roy. Soc. London A 144,* 75 (1934).
58. Weakley, T. J. R. *Structure and Bonding,* vol. 18 (West Berlin, 1974), Springer Verlag, p. 137.
59. Marcinkowska, K., S. Kaliaquine, and P. C. Roberge, *J. Catal. 90,* 49 (1984).
60. Ono, T., M. Anpo, and Y. Kubokawa, *J. Phys. Chem. 90,* 4780 (1986).
61. Thomas, R., E. M. van Oers, V. H. J. de Beer, and J. A. Moulijn, *J. Catal. 84,* 275 (1983).
62. Seyedmonir, S. R., and R. F. Howe, *J. Catal. 110,* 216 (1988).
63. Klemperer, W. G., V. V. Mainz, and R.-C. Wang, *Inorg. Chem. 24,* 1968 (1985).
64. Liu, Y. C., G. L. Griffin, S. S. Chan, and I. E. Wachs, *J. Catal. 94,* 108 (1985).
65. Cunningham, J., and B. K. Hodnett, *J. Chem. Soc. Faraday Trans. I 77,* 2777 (1981).
66. Djeghri, N., and S. J. Teichner, *J. Catal. 62,* 99 (1980).
67. Cundall, R. B., R. Rudman, and M. Salim, *J. Chem. Soc., Faraday Trans. I 72,* 1642 (1976).
68. Ramis, G., G. Busca, and V. Lorenzelli, *Zeit. Phys. Chemie. Neue Folge 153,* 189 (1987).
69. Ng, K. Y. S., and E. Gulari, *J. Catal. 92,* 340 (1985).
70. Shimada, H., J. Hiraishi, T. Sato, Y. Yoshimura, and A. Nishijima, *Proc.* 9th Int. Conf. Raman Spectrosc., Tokyo (1984), p. 734.
71. Quincy, R. B., M. Houalla, and D. M. Hercules, *J. Catal. 106,* 85 (1987).
72. Callahan, J. L., R. K. Grasselli, E. C. Millberger, and H. A. Strecker, *Ind. Eng. Chem. Prod. Res. Dev. 9,* 134 (1970).
73. Krenzke, L. D., and G. W. Keulks, *J. Catal. 61,* 316 (1980): ibid. *64,* 295 (1980).
74. Grasselli, R. K., and J. D. Burrington, *Adv. Catal. 30,* 133 (1981).
75. Grasselli, J. P., J. D. Burrington, and J. F. Brazdil, *Faraday Discuss. Chem. Soc. 72,* 203 (1981).
76. Glaeser, L. C., J. F. Brazdil, M. A. Hazle, M. Mehicic, and R. K. Grasselli, *J. Chem. Soc. Faraday Trans. 81,* 2903 (1985).
77. Hoefs, E. V., J. R. Monnier, and G. W. Keulks, *J. Catal. 57,* 331 (1971).
78. Grasselli, R. K., J. F. Brazdil, and J. D. Burrington, in *Proc. 8th Int. Congr. Catal.* vol. V, (1984), p. 369.
79. Matsuura, I., R. Schut, and K. Hirakawa, *J. Catal. 63,* 152 (1980).
80. Kovats, W. D., and C. G. Hill, Jr., *Appl. Spectrosc. 40,* 1215 (1986).
81. de Beer, V. H. J., M. J. M. van der Aalst, C. J. Machiels, and G. C. A. Schuit, *J. Catal. 43,* 78 (1976).
82. de Beer, V. H. J., T. H. M. van Sint Fiet, J. F. Engelen, A. C. van Haandel, M. W. J. Wolfs, C. H. Amberg, and G. C. A. Schuit, *J. Catal. 27,* 357 (1957).
83. Lopez Agudo, A., J. L. G. Fierro, C. Caceres, J. Laine, and F. Severino, *Appl. Catal. 30,* 185 (1987).
84. Caceres, C., J. L. G. Fierro, A. Lopez Agudo, F. Severino, and J. Laine, *J. Catal. 97,* 219 (1986).
85. Pecoraro, T. A., and R. R. Chianelli, *J. Catal. 67,* 430 (1981).
86. Singhal, G. H., R.L. Espino, and J. E. Sobel, *J. Catal. 67,* 446 (1981).
87. Schuit, G. C. A., and B. C. Gates, *AIChE J. 19,* 417 (1973).
88. Hagenbach, G., P. H. Courty, and B. Delmon, *J. Catal. 31,* 264 (1973).
89. Furminsky, F., and C. H. Amberg, *Can. J. Chem. 53,* 2542 (1975).

90. de Beer, V. H. J., J. G. J. Dahlmans, and J. G. U. Smeets, *J. Catal. 42,* 467 (1976).
91. Topsoe, H., B. S. Clausen, R. Candia, C. Wivel, and S. Morup, *J. Catal. 68,* 433 (1981).
92. Wivel, C., R. Candia, B. S. Clausen, S. Morup, and H. Topsoe, *J. Catal. 68,* 453 (1981).
93. Topsoe, H., B. S. Clausen, R. Candia, C. Wivel, and S. Morup, *Bull. Chim. Soc. Chim. Belg. 90,* 1189 (1981).
94. Chung, K. S., and F. E. Massoth, *J. Catal. 64,* 320 (1980).
95. Chin, R., and D. M. Hercules, *J. Phys. Chem, 86,* 360 (1982); ibid. *86,* 3079 (1982).
96. Makovsky, L. E., J. M. Stencel, F. R. Brown, R. E. Tischer, and S. S. Pollack, *J. Catal. 89,* 334 (1984).
97. LoJacono, M., A. Cimino, and G. C. A. Schuit, *Gazz. Chim. Ital. 103,* 1281 (1973).
98. Payen, E., M. C. Dhamelincourt, P. Dhamelincourt, J. Grimblot, and J. P. Bonnelle, *Appl. Spectrosc. 36,* 30 (1982).
99. Kasztelan, S., E. Payen, H. Toulhoat, J. Grimblot, and J. P. Bonnelle, *Polyhedron, 5,* 157 (1986).
100. Schrader, G. L., and C. P. Cheng, *J. Catal. 85,* 488 (1984).
101. Petrakis, L., and F. E. Kiviat, *J. Phys. Chem. 80,* 606 (1976).
102. Ratnasamy, P., and J. Fripiat, *Trans. Faraday Soc. 66,* 2897 (1970).
103. Schuman S. C., and H. Shalit, *Catal. Rev. 4,* 245 (1970).
104. Katzer, J. R., and R. Sivasubramanian, *Catal. Rev. 20,* 155 (1979).
105. Bak, T., L. Dziembaj, J. Haber, and J. Ziolkowski, *Zesz. Nauk ISCO, 20,* 35 (1972).
106. Dufresne, P., E. Payen, J. Grimblot, and J. P. Bonnelle, *J. Phys. Chem. 85,* 2344 (1981).
107. Sombret, B., P. Dhamelincourt, F. Wallert, A. C. Muller, M. Bouquet, and J. Grosmangin, *J. Raman Spectrosc. 9,* 291 (1980).
108. Chan, S. S., and I. E. Wachs, *J. Catal. 103,* 224 (1987).
109. Wells, A. F., *Structural Inorganic Chemistry,* 5th ed. (Oxford University Press, London, 1984).
110. Shalvoy, R. B., P. J. Reucroft and B. H. Davis, *J. Catal. 56,* 336 (1979).
111. Kasztelan, S., H. Toulhoat, J. Grimblot, and J. P. Bonnelle, *Appl. Catal. 13,* 127 (1984).
112. Bachelier, J., J. C. Duchet, and D. Cornet, *J. Catal. 87,* 283 (1984).
113. J. Bachelier, J., M. J. Tilliett, J. C. Duchet, and D. Cornet, *J. Catal. 87,* 292 (1984).
114. Stencel, J. M., L. E. Makovksy, J. R. Diehl, and T. A. Sarkus, *J. Catal. 95,* 414 (1985).
115. Massoth, F. E., *Advances in Catalysis,* vol. 27 (Academic Press, New York, 1978).
116. Lopez Agudo, A., J. L. G. Fierro, C. Caceres, J. Laine, and F. Severino, *Appl. Catal. 30,* 185 (1987).
117. Laine, J., F. Severino, and R. Golding, *J. Chem. Technol. Biotechnol. 34A,* 387 (1984).
118. Payen, E., S. Kasztelan, J. Grimblot, and J. P. Bonnelle, *J. Raman Spectrosc. 17,* 240 (1986).
119. Kantschewa, M., F. Delannay, H. Jeziorowski, E. Delgado, S. Eder, G. Ertl, and H. Knozinger, *J. Catal. 87,* 482 (1984).
120. Newsome. D. S., *Catal. Rev.—Sci. Eng. 21,* 275 (1980).
121. Murchinson, C. B., in *Chemical Uses of Molybdenum, Proc. Int. Conf.* ed. H. F. Barry and P. C. Mitchell, Climax Molybdenum Co., Ann Arbor, Michigan (1985), p. 197.
122. Nakamoto, N., *Infrared and Raman Spectra of Inorganic and Coordination Compounds* (Wiley: New York, 1978).
123. Weinstock, N., H. Schulze, and A. Muller, *J. Chem. Phys. 59,* 5063 (1973).
124. Beattie, I. R., and T. R. Gilson, *J. Chem. Soc. A,* 2322 (1969).
125. Hauck, J., and A. Fadini, *Z. Naturforsch. 25B,* 422 (1970).
126. Busey, R. H., and O. L. Keller, Jr., *J. Phys. Chem. 41,* 215 (1964).
127. Khanna, R. K., W. S. Brower, B. R. Guscott, and E. R. Lippincott, *J. Res. Nat. Bur. Stand. Sect. A, 72,* 81 (1968).
128. Liegeois-Duyckaerts, M., and P. Tarte, *Spectrochim. Acta 28A,* 2037 (1972).

129. Stencel, J. M., J. Springer, and E. Silberman, *J. Mol. Struct. 41,* 11 (1977).
130. Biloen, P., and G. T. Pott, *J. Catal. 30,* 169 (1973).
131. Thomas, R., M. C. Mittelmeijer-Hazeleger, F. P. J. M. Kerkhof, J. A. Moulijn, J. Medema, and V. H. J. de Beer, in *Chemical Uses of Molybdenum, Proc. Int. Conf.*, ed. H. F. Barry and P. C. Mitchell, Climax Molybdenum Co., Ann Arbor, Michigan (1985), p. 85.
132. Thomas, R., F. P. J. M. Kerkhof, J. A. Moulijn, J. Medema, and V. H. J. de Beer, *J. Catal. 61,*559 1980.
133 Thomas, R. V. H. J. de Beer, and J. A. Moulign, *Bull Soc. Chim. Belg. 90* 1349 (1981).
134. Iannibello, A., P. L. Villa, and S. Marengo, *Gazz. Chem. Ital. 109,* 521 (1979).
135. Griffith, W.P., and T. D. Wickins, *J. Chem. Soc. A,* 1087 (1966).
136. Griffith, W. P., and P. J. B. Lesniak, *J. Chem. Soc. A,* 1066 (1969).
137. Busey, R. H., and O. L. Keller, Jr., *J. Chem. Phys. 41,* 215 (1964).
138. Salvati, L., Jr., L. E. Makovsky, J. M. Stencel, F. R. Brown, and D. M. Hercules, *J. Phys. Chem. 85,* 3700 (1981).
139. Tiltarelli, P., A. Iannibello, and P. L. Villa, *J. Solid State Chem. 37,* 95 (1981).
140. Chan, S. S., I. E. Wachs, and L. L. Murrell, *J. Catal. 90,* 150 (1984).
141. Chan, S. S., I. E. Wachs, L. L. Murrell, and N. C. Dispenziere, *J. Catal. 92,* 1 (1985).
142. Stencel, J. M., L. E. Makovsky, J. R. Diehl, and T. A. Sarkus, *J. Raman Spectrosc. 15,* 282 (1984).
143. Chan, S. S., I. E. Wachs, L. L. Murrell, L. Wang, and W. K. Hall, *J. Phys. Chem. 88,* 5831 (1984).
144. Carver, J. C., I. E. Wachs, and L. L. Murrell, *J. Catal. 94,* 213 (1985).
145. Wachs, I. E., C.C. Chersich, and J. H. Hardenbergh, *Appl. Catal. 13,* 335 (1985).
146. Horsley, J. A., I. E. Wachs, J. M. Brown, G. H. Via, and F. D. Hardcastle, *J. Phys. Chem. 91,* 4015 (1987).
147. Knee, F., and R. A. Condrate, Sr., *J. Phys. Chem. Solids, 40,* 1145 (1979).
148. Okada, K., H. Morikawa, F. Marumo, and S. Iwai, *Acta Crystallogr. B31,* 1200 (1975).
149. Ahuja, S. D., M. D. Derrien, and J. F. Le Page, *Ind. Eng. Chem., Prod. Res. Rev. 9,* 272 (1970).
150. K. T. Ng, and D. M. Hercules, *J. Phys. Chem. 80,* 2094 (1976).
151. Wu, M., and D. M. Hercules, *J. Phys. Chem. 83,* 2003 (1974).
152. Stencel, J. M., R. E. Tischer, G. J. Steigel, J. R. Diehl, J. F. Paterline, and L. E. Makovsky, unpublished results.
153. de Vries, S. L. K. F., and G. T. Pott, *Rec. Trav. Chim. Pays-Bas 96,* M115 (1977).
154. Thomas, R., J. A. Moulign, and F. P. J. M. Kerkhof, *Rec. Trav. Chim. Pays-Bas 96,* M134 (1977).
155. Banks, R. L., *Fortschr. Chem. Forsch. 25,* 39 (1972).
156. Robinson, R. C., L. A. Frederickson, and R. L. Jacobsen, *Proc. Div. Refin. Am. Petr. Inst. 52,* 163 (1972).
157. Peri, J. B., *J. Catal. 52,* 144 (1978).
158. Wagstaff, N., and R. Prins, *J. Catal. 59,* 434 (1979).
159. Boelhouwer, C., and A. A. Olsthoorn, *J. Catal. 44,* 197 (1976).
160. Kerkhof, F. D. J. M., J. A. Moulijn, and R. Thomas, *J. Catal. 56,* 279 (1979).
161. Ulbricht, K., and H. Kriegsmann, *Z. Chem. 6,* 232 (1966).
162. Ulbricht, K., and H. Kriegsmann, *Z. Anorg. Allg. Chem. 358,* 193 (1968).
163. Woodward, L. A., and H. L. Roberts, *Trans. Faraday Soc. 52,* 615 (1956).
164. Yao, H. C., and M. Shelef, *J. Catal. 44,* 392 (1976).
165. Nakamura, R., F. Abe, and E. Echigoya, *Chem. Lett. Japan Chem Soc. 51* (1981).
166. Olsthoorn, A.A., and C. Boelhouwer, *J. Catal. 44,* 197 (1976).
167. Wang, L., and W. K. Hall, *J. Catal. 82,* 177 (1983).

168. Bond, G. C., and P. Konig, *J. Catal. 77,* 309 (1982).
169. Bond, G. C., and K. Bruckman, *Faraday Discuss. 72,* 235 (1981).
170. Gazior, M., I. Gazior, and B. Grzybowska, *Appl. Catal. 10,* 87 (1984).
171. Rozeboom, F., M. C. Mittelmeijer-Hazeleger, J. A. Moulijn, J. Medema, J. H. J. de Beer, and P. J. Gellings, *J. Phys. Chem. 84,* 2783 (1980).
172. Bond, G. C., J. Sarkany, and G. D. Parfitt, *J. Catal. 57,* 476 (1979).
173. Dyrek, K., E. Servicka, and B. Grzybowska, *React. Kinet. Catal. Lett. 10,* 93 (1979).
174. Kozlowski, R., R. F. Pettifer, and J. M. Thomas, *J. Phys. Chem. 87,* 5176 (1983).
175. Saleh, R. Y., I. E. Wachs, S. S. Chan, and C. C. Chersich, *J. Catal. 98,* 102 (1986).
176. Wachs, I. E., S. S. Chan, and R. Y. Saleh, *J. Catal. 91,* 366 (1985).
177. Wachs, I. E., S. S. Chan, C. C. Chersich, and R. Y. Saleh, *Appl. Catal. 15,* 339 (1985).
178. Davis, B. H., *Appl. Surf. Sci. 19,* 200 (1984).
179. Freeman, G. B., D. J. Collins, J. C. Waters, and B. H. Davis, *J. Phys. Chem. 84,* 55 (1980).
180. Krebs, M. A., and R. A. Condrate, Sr., *J. Am. Cer. Soc. 65,* C144 (1982).
181. Anastossakis, E., B. Papanicolaou, and I. M. Asher, *J. Phys. Chem. Solids 36,* 667 (1975).
182. Simpson, S., and B. H. Davis, *J. Phys. Chem. 91,* 5664 (1987).
183. Stencel, J. M., E. B. Bradley, and F. R. Brown, *Appl. Spectrosc. 34,* 319 (1980).
184. Iannibello, A., S. Marengo, P. Tittarelli, G. Morelli, and A. Zecchina, *J. Chem. Soc., Faraday Trans. I 80,* 2209 (1984).
185. Doyle, W. P., and P. Eddy, *Spectrochim. Acta 23A,* 1903 (1967).
186. Grant, W. B., and S. Rodhokrishna, *Solid State Commun., 13,* 109 (1973).
187. Scheuerman, W., and G. J. Ritter, *J. Mol. Struct. 6,* 240 (1970).
188. Srivastova, R., and L. L. Chase, *Solid State Commun. 11,* 349 (1972).
189. Hart, T. R., R. L. Aggarwal, and B. Lax, in *Proc. 2nd Int. Conf. Light Scattering Solids,* Boston, MA (1971), p. 174.
190. Ono, T., T. Yamanaka, Y. Kubokawa, and M. Komiyana, *J. Catal. 109,* 423 (1988).
191. Scott, J. F., *J. Chem. Phys. 53,* 852 (1970).
192. Miller, P. J., and C. A. Cody, *Spectrochim. Acta 38A,* 555 (1982).
193. Jansen, V. M., J. Pebler, and K. Dehn Nicke, *Z. Anorg. Allg. Chem. 495,* 120 (1982).
194. Chan, S. S., and A. T. Bell, *J. Catal. 89,* 433 (1984).
195. Ryndin, Y. A., R. F. Hicks, A. T. Bell, and Y. I. Yermakov, *J. Catal. 70,* 287 (1981).
196. Poels, E. K., E. H. van Brokehoeven, W. A. A. van Barneveld, and V. Ponec, *React. Kinet. Catal. Lett. 18,* 233 (1981).
197. Boldish, S. I., and W. B. White, *Spectrochim Acta 35,* 1235 (1979).
198. Gopinath, C. R., and I. D. Brown, *J. Raman Spectrosc. 12,* 278 (1982).
199. Swanson, B. I., C. Machell, G. W. Beall, and W. O. Milligan, *J. Inorg, Nucl. Chem. 40,* 694 (1978).
200. Bosworth, Y. M., and R. J. H. Clark, *Inorg. Chem. 14,* 170 (1975).
201. Papatheodorou, G. M., and M. A. Capote, *J. Chem. Phys. 69,* 2067 (1978).
202. Clark, R. J. H., and W. R. Trumble, *Inorg. Chem. 15,* 1030 (1976).
203. Reichmann, M. G., and A. T. Bell, *Langmuir 3,* 111 (1987).
204. Gratzel, M., and F. P. Rotzinger, *Inorg. Chem. 24,* 2321 (1985).
205. Karagounis, G., and R. Issa, *Nature, 195,* 1196 (1962).
206. Karagounis, G., and R. Issa, *Z. Elektrochem, 66,* 874 (1962).
207. Pershina, E. V., and Sh. Sh. Raskin, *Dokl. Akad. Nauk S.S.S.R. 150,* 1022 (1963).
208. Pershina, E. V., and Sh. Sh. Raskin, *Optics Spectr., Supl. 3,* 164 (1963).
209. Hendra, P. J., and E. J. Loader, *Nature 216,* 789 (1967).
210. Hendra, P. J., and E. J. Loader, *Nature 217,* 637 (1968).
211. Hendra, P. J., and E. J. Loader, *Trans. Faraday Soc. 67,* 828 (1971).
212. Hendra, P. J., E. J. Horder, and E. J. Loader, *Chem. Commun.* 563 (1970).

213. Loader, E. J., *J. Catal. 22,* 41 (1971).
214. Hendra, P. J., I. D. M. Turner, E. J. Loader, and M. Stacy, *J. Phys. Chem. 78,* 300 (1974).
215. Hendra, P. J., J. R. Horder and E. J. Loaer, *J. Chem. Soc.* 1766 (1971).
216. Basila, M. R., T. R. Kantner, and K. H. Rhee, *J. Phys. Chem. 68,* 3197 (1964).
217. Basila, M. R., and T. R. Kantner, *J. Chem. Phys. 70,* 1681 (1966).
218. Schwarz, J. A., *J. Vac. Sci. Technol. 12,* 321 (1975).
219. Hendra, P. J., *SPEX Speaker 24,* (1974).
220. Winde, H., and V. Denine, *Z. fur Phys. Chem. 244,* 283 (1970).
221. Winde, H., and V. Denine, *Z. fur Chem. 10,* 64 (1970).
222. Turner, I. D. M., Ph.D. thesis, Southampton (1974).
223. Hair, M. L., *Infrared Spectroscopy in Surface Chemistry* (Dekker, New York, 1967).
224. Little, L. H., *Infrared Spectra of Adsorbed Species* (Academic Press, London, 1966).
225. R. P. Cooney, G. Curthoys, and N. T. Tam, *Advan. Catal. 24,* 293 (1975).
226. Egerton, T. A., and A. H. Hardin, *Catal. Rev. Sci. Eng. 11,* 71 (1975).
227. Egerton, T. A., A. H. Hardin, Y. Kozirovski, and N. Sheppard, *J. Catal. 32,* 343 (1974).
228. Tam, N. T., R. P. Cooney, and G. Curthoys, *J. Colloid Interface Sci. 51,* 340 (1975).
229. Bahnick, D. A., and W. B. Person, *J. Chem. Phys. 48,* 1251 (1968).
230. Jeziorowski, H., and H. Knozinger, *Chem. Phys. Lett. 43,* 37 (1976).
231. Angell, C. L., *J. Phys. Chem. 77,* 222 (1973).
232. Tam, N. T., R. P. Cooney, and G. Curthoys, *J. Chem. Soc., Faraday I 72,* 2577 (1976).
233. Tam, N. T., R. P. Cooney, and G. Curthoys, *J. Chem. Soc. Faraday I 72,* 2592 (1976).
234. Herzberg, G. *Infrared and Raman Spectra of Polyatomic Molecules* (Van Nostrand, New York, 1945).
235. Hardin, A. H., M. Klemes, and B. A. Morrow, *J. Catal. 62,* 316 (1980).
236. Egerton, T. A., A. H. Hardin, and N. Sheppard, *Can. J. Chem. 54,* 586 (1976).
237. Ward, J. W., *J. Catal. 10,* 34 (1968).
238. Cooney, R. P., and N. T. Tam, *Aust. J. Chem. 29,* 507 (1976).
239. Freeman, J. J., and M. L. Unland, *J. Catal. 54,* 183 (1978).
240. Murray, C. A., and T. J. Greytak, *J. Chem. Phys. 71,* 3355 (1979).
241. Zundel, G., in *The Hydrogen Bond,* ed. P. Schuster, G. Zundel, and C. Sandorfy (North Holland, Amsterdam, 1976).
242. Knozinger, H., in *The Hydrogen Bond,* ed. P. Schuster, G. Zundel and C. Sandorfy (North Holland, Amsterdam, 1976).
243. Peri, J. B., *J. Phys. Chem. 70,* 2937 (1966).
244. Zhang, H. B., and G. L. Schrader, *J. Catal. 99,* 461 (1986).
245. Schrader, G. L., and C. P. Cheng, *J. Phys. Chem. 87,* 3675 (1983).
246. Mone, R., in *Preparation of Catalysts,* ed. B. Delman, P. A. Jacobs, and G. Poncelet (Elsevier, Amsterdam, 1976).
247. Segawa, K. I., and W. K. Hall, *J. Catal. 76,* 133 (1982).
248. Ismail, H, M., C. R. Theocharis, D. N. Waters, M. I. Zaki, and R. B. Fahim, *J. Chem. Soc., Faraday Trans. I 83,* 1601 (1987).
249. Kraser, W., A. Lanade, and E. Koglin, *J. Raman Spectrosc. 6,* 209 (1977).
250. W. Kiefer, and H. J. Bernstein, *Appl. Spectrosc. 25,* 609 (1971).
251. Kiefer, W., and H. J. Bernstein, *Mol. Phys. 23,* 835 (1972).
252. Eischens, R. P., S. A. Francis, and W. A. Pliskin, *J. Phys. Chem. 60,* 194 (1956).
253. Krasser, W., A. Fadini, A. J. Renouprez, *J. Catal. 62,* 94 (1980).
254. Stencel, J. M., and E. B. Bradley, *Rev. Sci. Instr. 49,* 1163 (1978).
255. Stencel, J. M., and E. B. Bradley, *J. Raman Spectrosc. 8,* 203 (1979).
256. Bradley, E. B., and K. A. Arunkumar, *Spectrosc. Lett. 15,* 113 (1982).
257. Marzouk, H. A., K. A. Arunkumar, and E. B. Bradley, *Surf. Sci. 147,* 477 (1984).

258. Bradley, E. B., K. A. Arunkumar, and J. M. Stencel, *Spectral Line Shapes,* vol. 2 (Walter H. Gruyter & Co., Berlin–New York, 1983), p. 625.
259. Arunkumar, K. A., H. A. Marzouk, and E. B. Bradley, *Applications of Laser Chem. and Diagnostics (SPIE) 482,* 130 (1984).
260. Richardson, N. V., and A. M. Bradshaw, *Surf. Sci. 88,* 255 (1979).
261. Anderson, S., *Solid State Commun. 21,* 75 (1977).
262. Mitchell, G. E., J. L. Gland, and J. M. White, *Surf. Sci. 131,* 167 (1983).
263. Marzouk, H. A., E. B. Bradley, and K. A. Arunkumar, *Spectrosc. Lett. 18,* 189 (1985).
264. Vannice, M. A., *Catal. Rev. Sci. Eng. 14,* 153 (1976).
265. Arunkumar, K. A., H. A. Marzouk, and E. B. Bradley, *Rev. Sci. Instrum. 55,* 905 (1984).
266. Greenler, R. G., and T. L. Slager, *Spectrochim. Acta, Part A 29,* 193 (1973).
267. Campion, A., and D. R. Mullins, *Chem. Phys. Lett. 94,* 576 (1983).
268. Campion, A., J. K. Brown, and V. M. Grizzle, *Surf. Sci., 115,* L153 (1982).
269. Campion, A., and D. R. Mullins, *Surf. Sci. 158,* 263 (1985).
270. Mullins, D. R., and A. Campion, *Chem. Phys. Lett. 110,* 561 (1984).
271. Campion, A., *J. Vac. Sci. Technol. B3,* 1404 (1985).
272. Campion, A., *Ann. Rev. Phys. Chem. 36,* 549 (1985).
273. Hallmark, V. M., "Surface Raman Spectroscopy: A New Adsorbate Vibrational Probe and Its Selection Rules." Ph.D. dissertation, University of Texas at Austin (1984).
274. Mullins, D. R., "Surface Raman Spectroscopy: Investigations on Smooth, Stepped and Kinked Crystal Faces and Angular Resolution Studies." Ph.D. dissertation, University of Texas at Austin (1984).
275. Wu, J. A., H. Metiu, and B. Kirtman, *Surf. Sci. 177,* 101 (1986).
276. Chou, C. C., C. E. Reed, J. C. Hemminger, and S. Ushioda, *J. Electron Spectrosc. and Related Phenom. 29,* 401 (1983).
277. Udagawa, M., C. C. Chou, J. C. Heminger, and S. Ushioda, *Phys. Rev. B,* 6843 (1981).
278. Furtak, T. E., and J. Kester, *Phys. Rev. Lett. 45,* 1652 (1980).
279. Yamada, Y. and Y. Yamamoto, *Chem. Phys. Lett. 77,* 520 (1981).
280. Fleischman, M., P. R. Graves, I. R. Hilland, and J. Robinson, *Chem. Phys. Lett. 95,* 322 (1983).
281. Parker, W. L., A. R. Siedle, and R. M. Hexter, Technical Report 9, ONR, Contract N00014-77-C-0209.
282. Fleischman, M., and I. R. Hill, in *Surface Enhanced Raman Scattering,* ed. R. K. Chang and T. E. Furtak (Plenum Press, New York 1981).
283. Parker, W. L., A. R. Siedle, and R. M. Hexter, Tech. Report 10, ONR, Contract N00014-77-C-0209.
284. Parker, W. L., A. R. Siedle, and R. M. Hexter, *J. Am. Chem. Soc. 107,* 264 (1985).
285. Hexter, R. M., and M. G. Albrecht, *Spectrochim. Acta 35A,* 233 (1979).
286. Zhang, H. B., and G. L. Schrader, *J. Catal. 95,* 325 (1985).
287. Lokten, S. M., A. N. Bashkirov, E. V. Slivinskii, L. I. Zvezdkina, and Yu B. Kagan, *Kinet. Catal. 14,* 175 (1973).
288. King, D. L., and J. B. Peri, *J. Catal. 79,* 164 (1983).
289. *Chem. Eng. News* (Feb. 17, 1986).
290. Weisz, P. B., *Chemtech 3,* 498 (1973).
291. Haynes, H. W., *Catal. Rev. 17,* 273 (1978).
292. Rabo, J. A., ed., *Zeolite Chemistry and Catalysis,* ACS Monograph 171 (American Chemical Society, Washington, D.C., 1976).
293. Katzer, J. R., ed., *Molecular Sieves—II,* ACS Symposium Series 40 (American Chemical Society, Washington, DC, 1977).

294. Breck, D. W., *Zeolite Molecular Sieves—Structure, Chemistry, and Use* (Wiley, New York, 1974).
295. Meier, W. M., and J. B. Uytterhoeven, eds., *Molecular Sieves, Advance in Chem. Series 121* (American Chemical Society, Washington, DC, 1973).
296. Stucky, G. D. and F. G. Dwyer, ed., *Zeolite Chemistry,* ACS Symposium Series 218, (American Chemical Society, Washington, DC, 1983).
297. Venuto, P. B., and E. T. Habib, in *Fluid Catalytic Cracking with Zeolite Catalysts,* M. Dekker, ed. (New York, 1979).
298. Barrer, R. M., *Zeolites 1,* 130 (1981).
299. Sand, L. B., *Pure Appl. Chem. 52,* 2105 (1980).
300. Culfaz, A., and L. B. Sand, *Advan. Chem. Series 121,* 140 (1973).
301. McNicol, B. D., G. T. Pott, and K. R. Lloss, *J. Phys. Chem. 76,* 3388 (1972).
302. Barrer, R. M., J. W. Baynham, F. W. Bultitude, and J. Meier, *J. Chem. Soc.,* 195 (1959).
303. Kerr, G. T., *J. Phys. Chem. 70,* 1047 (1966).
304. Zhdanov, S. P., *Advan. Chem. Series 101,* 20 (1971).
305. Flanigen, E. M., *Advan. Chem. Series 121,* 119 (1973).
306. Aiello, R., R. M. Barrer, and I. S. Kerr, *Advan. Chem. Series 101,* 44 (1971).
307. Angell, C. L., *J. Phys. Chem. 77,* 222 (1973).
308. Angell, C. L., and P. C. Schaffer, *J. Phys. Chem. 69,* 3463 (1965).
309. Hendra, P. J., J. R. Horder, and E. J. Loader, *J. Chem. Soc. A,* 1766 (1971).
310. Porto, S. P. S., and R. S. Krishnan, *J. Chem. Phys. 47,* 1009 (1967).
311. Hendra, P. J., and E. J. Loader, *Trans. Faraday Soc. 67,* 828 (1971).
312. Egerton, T. A., A, H. Hardin, Y. Kosirovski, and N. Sheppard, *Chem. Commun.,* 888 (1971).
313. Stroud, C., J. M. Stencel, and L. T. Todd, *J. Phys. Chem. 83,* 2378 (1979).
314. Badrinarayan, M. K., J. M. Stencel, and L. T. Todd, *J. Phys. Chem. 84,* 456 (1980).
315. Bates, J. B., *J. Chem. Phys. 56,* 1910 (1972).
316. Flanigen, E. M., H. Khatami, and H. A. Szymanski, *Advan. Chem. Series 101,* 201 (1971).
317. Guth, J. L., P. Gaullet, P. Jacques, and R. Wey, *Bull. Soc. Chim. France 3–4,* 121 (1980).
318. Guth, J. L., P. Gaullet, and R. Wey, in *Proc. 5th Int. Conf. on Zeolites,* ed. L. V. C. Rees (Heyden London), p. 30.
319. Ueda, S., N. Kageyama, and M. Koizumi, in *Proc. 6th Int. Conf. on Zeolites,* Reno, NV (USA), (July 10–15 1983).
320. Engelhardt, G., D. Hoebbel, M. Tarmak, A. Samoson, and E. Lippmaa, *Z. anorg. allg. Chem. 484,* 22 (1982).
321. Angell, C.L., and W. H. Flank, in *Molecular Sieves II,* ed. J. R. Katzer, ACS Symposium Series, *40,* 194 (1977).
322. Roozeboom, F., H. E. Robson, and S. S. Chan, *Zeolites 3,* 321 (1983).
323. Moolenaar, R. J., J. C. Evans, and L. D. McKeever, *J. Phys. Chem. 74,* 3629 (1970).
324. Dutta, P. K., and B. D. Barco, *J. Phys. Chem. 89,* 1861 (1985).
325. Dutta, P. K., and B. D. Barco, *Chem. Soc. Chem. Commun.,* 1297 (1985).
326. Dutta, B. D. Barco, and D. C. Shieh, *Chem. Phys. Lett. 127,* 200 (1986).
327. Dutta, P. K., D. C. Shieh, and M. Puri, *J. Phys. Chem. 91,* 2332 (1987).
328. de Kander, J. J. P. M., I. E. Maxwell, and P. J. Trotter, *J. Chem. Soc. Chem. Commun.,* 773 (1972).
329. Fateley, W. G., F. R. Dollish, N. T. McDevitt, and F. F. Bentley, *Infrared and Raman Selection Rules for Molecular and Lattice Vibrations* (Wiley-Interscience, New York 1972).
330. Ariai, J., and S. R. P. Smithe, *J. Phys. C. 14,* 1193 (1981).
331. Matson, D. W., S. K. Sharma, and J. A. Philpotts, *J. Non-Cryst. Solids 58,* 323 (1983).
332. McMillan, P., B. Piriou, and A. Navrotsky, *Geochem. Geophys. Acta 46,* 2021 (1982).
333. McMillan, P., *Am. Mineral. 69,* 662 (1984).

334. Henderson, C. M. B., and D. Taylor, *Spectrochim. Acta, Part A 33,* 283 (1977).
335. Bellamy, L. J., *The Infrared Spectra of Complex Molecules* (Chapman and Hall, London, 1975).
336. Matson, D. W., S. K. Sharma, and J. A. Philpotts, *J. Non-Cryst. Solids 58,* 323 (1983).
337. McKeown, D. A., F. L. Galeener and G. E. Brown, *J. Non-Cryst. Solids 68,* 361 (1984).
338. Dutta, P. K., and D. C. Shieh, *J. Phys. Chem. 90,* 2331 (1986).
339. Dutta, P. K., and D. C. Shieh, *Appl. Spectrosc. 39,* 343 (1985).
340. Sharma, S. K., J. A. Philpotts, and D. W. Matson, *J. Non-Cryst. Solids 71,* 403 (1985).
341. Griffith, W. P., *J. Chem. Soc. A,* 1372 (1969).
342. Galeener, F. L., *Solid State Commun. 44,* 1037 (1982).
343. Hayachi, S., K. Suzuki, S. Shin, K. Hayomizu, and O. Yamamoto, *Chem. Phys. Lett. 113,* 368 (1985).
344. Baranska, H., B. Czerwinska, and A. Labudzinska, *J. Mol. Struct. 143,* 485 (1986).
345. Miecznikowski, A., and J. Hanuza, *Zeolites 7,* 249 (1987).
346. Andersson, S.L. T., S. T. Lundin, S. Jaras, and J. E. Otterstadt, *Appl. Catal. 9,* 317 (1984).
347. Occelli, M. L., D. Psaras, and S. L. Suib, *J. Catal. 96,* 363 (1985).
348. Wormsbecher, R. F., A. W. Peters and J. M. Maselli, *J. Catal. 100,* 130 (1986).
349. Occelli, M. L., and J. M. Stencel, Reprint, Int. Symp. on Zeolites as Catalysts, Wurzburg, FDR (1988).
350. McKay, D. L., U.S. Patent No. 4,153,536 (1979).
351. Readal, T. C., J. D. McKinney, and R. A. Titmus, U.S. Patent No. 3,977,963 (1976).
352. Mitchell, B. R., and H. E. Swift, U.S. Patent No. 4,101,417 (1978).
353. Briel, R. F., U.S. Patent No. 2,901,419 (1959).
354. Berkus, B. J., and D. K. McKay, U.S. Patent No. 4,377,494 (1983).
355. Mathy, E. V., and C. J. Helmers, U.S. Patent No. 2,564,268 (1951).
356. Occelli, M. L., and J. V. Kennedy, U.S. Patent No. 4,465,588 (1984).
357. Occelli, M. L., and H. E. Swift, U.S. Patent No. 4,466,884 (1984).
358. Occelli, M. L., G.B. Patent No. 2,145,345A (1985).
359. Occelli, M. L., D. C. Kowalczyk, and C. L. Kibby, *J. Appl. Catal. 16,* 227 (1985).
360. Grim, R. E., *Clay Mineralogy* (McGraw Hill, New York, 1968).
361. Occelli, M. L., and J. M. Stencel, Reprint, Symp. on Advances in Catalytic Cracking, ACS, New Orleans, LA, (Aug. 30–Sept. 4, 1988).
362. Occelli, M. L., Reprint Symp. on Advances in Catalytic Cracking, ACS, New Orleans, LA (Aug. 30–Sept. 4, 1988).
363. Griffith, W. P., and P. J. B. Lesniak, *J. chem. Soc. A,* 1066 (1969).
364. Griffith, W. P., and T. D. Wickins, *J. Chem. Soc. A,* 1087 (1966).
365. Hanuza, J., K. Hermanowicz, W. Oganowski, and B. Trzebiatowska, *Bull. Pol. Acad. Sci. 31,* 139(1984).
366. Brown, R. G., and S. D. Ross, *Spectrochim. Acta 28A,* 1263 (1972).
367. Johnson, W. L., *Microbeam Anal.,* 26 (1986).
368. Tunistra, F., and J. L. Koenig, *J. Chem. Phys. 53,* 1126 (1970).
369. Nakamizo, M., R. Kammereck, and P. L. Walker, Jr., *Carbon 12,* 259 (1974).
370. Nakamizo, M., H. Honda, and M. Inagaki, *Carbon 16,* 281 (1978).
371. Vidana, R. P., B. D. Fischbach, L. J. Willis, and R. M. Loehr, *Solid State Commun. 39,* 341 (1981).
372. Tuinstra, F., and J. L. Koenig, *J. Compos. Material 4,* 492 (1970).
373. Friedal, R. A., and G. L. Carlson, *Chem. Ind. (London) 40,* 1128 (1971).
374. Nathan, M. I., J. E. Smith, and K. N. Tu, *J. Appl. Phys. 45,* 2370 (1974).
375. Vidano, R., and B. D. Fischbach, *J. Am. Cer. Soc. 61,* 13 (1978).
376. Mernaugh, T. P., R. P. Cooney, and J. Spink, *J. Raman Spectrosc. 16,* 57 (1985).

377. Kim, J.-J., *Chem. Phys. Lett. 122,* 139 (1985).
378. Tsang, J. C., J. E. Demuth, P. N. Sanda, and J. R. Kirtley, *Chem. Phys. Lett. 76,* 54 (1980).
379. Elman, B. S., M. S. Dresselhaus, G. Dresselhaus, E. W. Maby, and H. Mazurek, *Phys. Rev. B24,* 1027 (1981).
380. Ishida, H., H. Fukuda, G. Katagiri, and A. Ishitani, *Appl. Spectrosc. 40,* 322 (1986).
381. Shirakawa, H., T. Ito, and S. Ikeda, *J. Polymers 4,* 460 (1973).
382. Baruya A., *Macromolecules 16,* 578 (1983).
383. Johnson, C. A., and K. M. Thomas, *Fuel 63,* 1073 (1984).
384. Baker, R. T. K., and P. S. Harris, *Chemistry and Physics of Carbon,* ed. P. L. Walker (Dekker, New York, 1978), p. 83.
385. Baker, R. T. K., *Catal. Rev. Sci. Eng. 19,* 161 (1979).
386. Lespade, P., R. Al Jishi, and M. S. Dresselhaus, *Carbon 20,* 427 (1982).
387. Espinat, D., H. Dexpert, E. Freund, G. Martino, M. Couzi, P. Lespade, and F. Cruege, *Appl. Catal. 16,* 343 (1985)

5 Surface Enhanced Raman Spectroscopy

INTRODUCTION

The recognition of extraordinary enhancements in Raman intensities from molecules adsorbed on certain noble metals, discussed in only a few papers published in the period 1974 to 1977 (1–3), has caused an avalanche of experimental Raman activity in the last dozen years. The pace of this activity is indicated by the fact that the enhancement factor of 10^6 for Raman intensities was discussed for the first time in 1977; only six years later more than 350 papers and one book on surface enhanced Raman spectroscopy (SERS) already had been published. By the middle of 1985 the number of published papers had jumped to over 570 (4), and since then it has continued to increase at a steady, rapid pace. This plethora of data on SERS has been dedicated primarily to defining materials and experimental conditions under which SERS can be observed and theoretical reasons for the observed effect. However, as will be discussed in this chapter, these SERS data also provide a solid basis for future development of its potential for catalysis. Only recently has SERS been applied to the study of catalysts, opening the door to further exploitation of its in situ capability for the study of solid–liquid and solid–gas interfaces at high pressures.

Differences between SERS and normal Raman (NR) spectroscopy and other surface techniques were summarized in Table 3.1. The sensitivity of SERS, as low as 10^{-3} monolayer, is substantially better than that of NR spectroscopy. As with NR spectroscopy, this sensitivity can be used under conditions that are near those for actual process conditions. The use of SERS for the study of all metal and metal oxide surfaces has not proved possible; Cu, Ag, and Au have been studied the most. But, the number of metals that have been shown to promote surface enhanced spectra (SES) now include Ni, Pt, Pd, Rh, Ti, Li, Na, K, Cd, In, Co, and Hg. This list is not so large as the list of materials that can be investigated by NR spectroscopic techniques. However, NR spectroscopy cannot always be easily applied to the study of adsorbates or surfaces, and it also requires that investigators use "tricks of the trade" to maximize intensities while using a minimal amount of laser power to observe relatively small concentrations of molecular entities. For SERS, the increase in the number of metals shown to

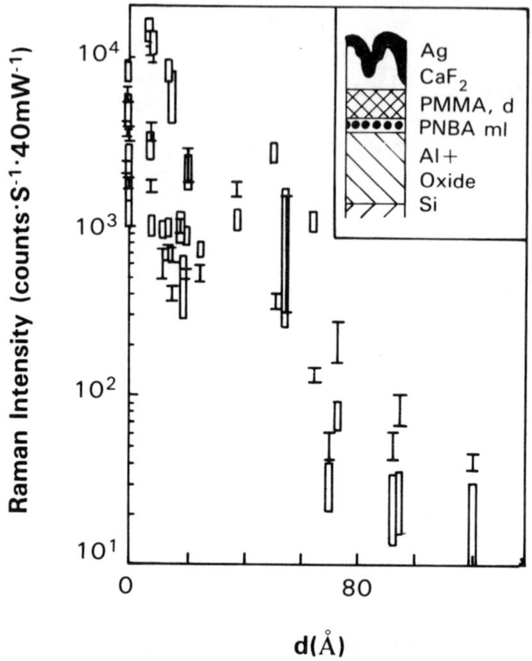

Figure 5.1. Raman intensity data as a function of the polymethylmethacrylate (PMMA) spacer thickness for the 1595 cm^{-1} (open symbols) and 1130 cm^{-1} (closed symbols) bands of *p*-nitrobenzoic acid (PNBA). Silicon substrate, aluminum, ~ 230 nm, oxide ~ 3 nm, PNBA monolayer, PMMA spacer ~ *d* nm, CaF_2 ~ 40 nm, Ag ~ 20 nm (5).

generate enhanced Raman band intensities is the result of a more detailed understanding of the origin of such surface enhancements and an increased awareness of the fundamental data obtained for adsorption of organic molecules on Au, Cu, and Ag metal surfaces.

An important aspect of SERS is that it produces enhancements only for molecules in direct interaction with the surface. For example, the first adsorbed monolayer may produce scattering enhancements of 10^3 to 10^6, whereas intensities for molecules located in the bilayer or farther from the surface (~1.0 nm) will be about two orders of magnitude smaller (for example, see Figure 5.1). Theoretically, such a rapid decrease in scattering cross sections as a function of bonding distances would be readily explainable if the basis for SERS were purely chemical in nature. Such a dependency of intensities on bonding distances implies that SERS may become an extremely sensitive in situ probe for chemisorption. In addition, a 10^3 to 10^6 enhancement in Raman intensities has been observed to occur for molecules adsorbed at fractions of a monolayer, beyond which intensities of the Raman scattered radiation may be unchanged (see below). The details of such sites, for example, whether they could be involved in catalytic reactions, and the commonality of such observations to other systems exhibiting SERS, have yet to be thoroughly addressed.

The bands in a surface enhanced spectrum (SES) normally are highly depolarized, a large number of otherwise Raman inactive modes may be activated, and the relative intensities of the bands are not thoroughly understood. These complications, however, are expected to become less important as properties of surfaces necessary to produce SERS are studied more extensively.

Two general conditions are required for SERS to occur: (1) a specific surface morphology having roughness within a certain size range is needed, and (2) suitable dielectric properties of the adsorbent are required. The surface roughness that seems to be required for the observation of SERS deserves special attention. This size seems to be on the order of atomic dimensions, but all such sites are not necessarily equal in the creation of SES (5–7). Classical electrodynamics has been successfully used to model the influence of roughness on SERS intensities. In the case of modeling, surfaces containing arrays of isolated and small ($\sim$10 nm) particles have been calculated to produce enhancements that coincide with experimental observations. In addition, the size of these particles is of interest in catalysis. Smaller-sized particles may not be suitable for the generation of as intense SERS effects because very small particle sizes increase the chances that conduction electrons will suffer energy losses in collisions with the surface of the particle. Radiation damping of the SERS effect also can occur, at particle sizes greater than approximately 200 nm. Relative to the dielectric constant criterion, to sustain high local electric fields and field gradients materials must have a small imaginary term in the complex dielectric. The metals Au, Ag and Cu fit this category, with Ag having low losses (high reflectivity) throughout the visible spectrum and Au and Cu having high reflectivity for wavelengths in the red end of the visible spectrum.

Other features of SERS that make it attractive for the study of catalysts include the relatively high resolution obtainable when Raman spectrometers are used, where the spectral resolution is normally limited only by the intrinsic width of the vibrational modes of the species under investigation. The low frequency region of the Raman spectra, close to the exciting radiation, also can be monitored. Observation of these modes implies the detection of external types of vibrations, in which an adsorbed molecule vibrates relative to the surface.

Silver was the first metal on which SERS data were obtained. This early use, coupled with the fact that Ag surfaces have been found to cause SES for adsorbed molecules using a broad range of visible excitation frequencies, has produced a preponderance of information on SERS using Ag substrates over the last decade. Hence, the following discussion is organized in terms of the substrates used, Ag, Cu, Au, as well as combinations of these metals with other metals, and finally other metals by themselves. Only the part of the current SERS literature that is related to catalysis is discussed. A comprehensive listing of publications and books on SERS was given recently by Seki (4).

Ag SURFACES

Much of the research on the SERS effect has been associated with describing and understanding data of pyridine adsorbed on Ag electrodes. Instead of reviewing this literature, (reviews are presented in references 5 and 8–10), this section will focus on adsorbates and modified adsorbent surfaces that produce SERS and that relate to catalytic systems. A considerable amount of the research mentioned in the reviews (5, 8–10) discusses the criteria for acquiring SERS data; in general, one important criterion is the observation of strong vibrational intensities from pyridine when it is adsorbed on a surface.

Organic Molecules

Moody, Vo-Dinh, and Fletcher (11) used discretely sized polystyrene latex spheres on which Ag layers of different thicknesses were vapor-deposited. These latex spheres allowed approximate control of the size of the light scattering particle, while vapor deposition of Ag enabled close control of the Ag thickness. Benzopyrene and 1-nitropyrene were adsorbed onto these surfaces and the SERS intensities monitored as a function of particle size, Ag layer thickness, and laser excitation frequency. Figure 5.2 shows some of the results. Although a change in the macroscopic size of these latex particles provided an adjustable parameter that could be used to maximize Raman intensities for very small quantities of adsorbate, the roughness of the Ag overlayer was not exactly quantified.

Competitive adsorption of d_5-pyridine and 3-methylpyridine on an Ag electrode was investigated by Ikezawa et al. (12). Although this work was performed at a electrode potential of -0.6 V (SCE), the data showed that coadsorption behavior could be approximated by Langmuir isotherms for a binary system, and that steric effects associated with 3-methylpyridine adsorption strongly inhibited the subsequent adsorption of pyridine.

The adsorption of pyridine on silica and alumina surfaces has been shown by NR and infrared spectroscopic investigations to occur through the lone electron pair of the nitrogen atom interacting with Lewis or Bronsted acid sites or by H-bonding with surface hydroxyl sites. The influence of the nitrogen lone electron pair during adsorption of pyridine onto metals exhibiting SERS effects also has been shown to be important. This bonding shifts vibrational modes of the adsorbed pyridine away from their corresponding positions in the liquid state. Recently, the relative amount of this shift for the different modes of pyridazine

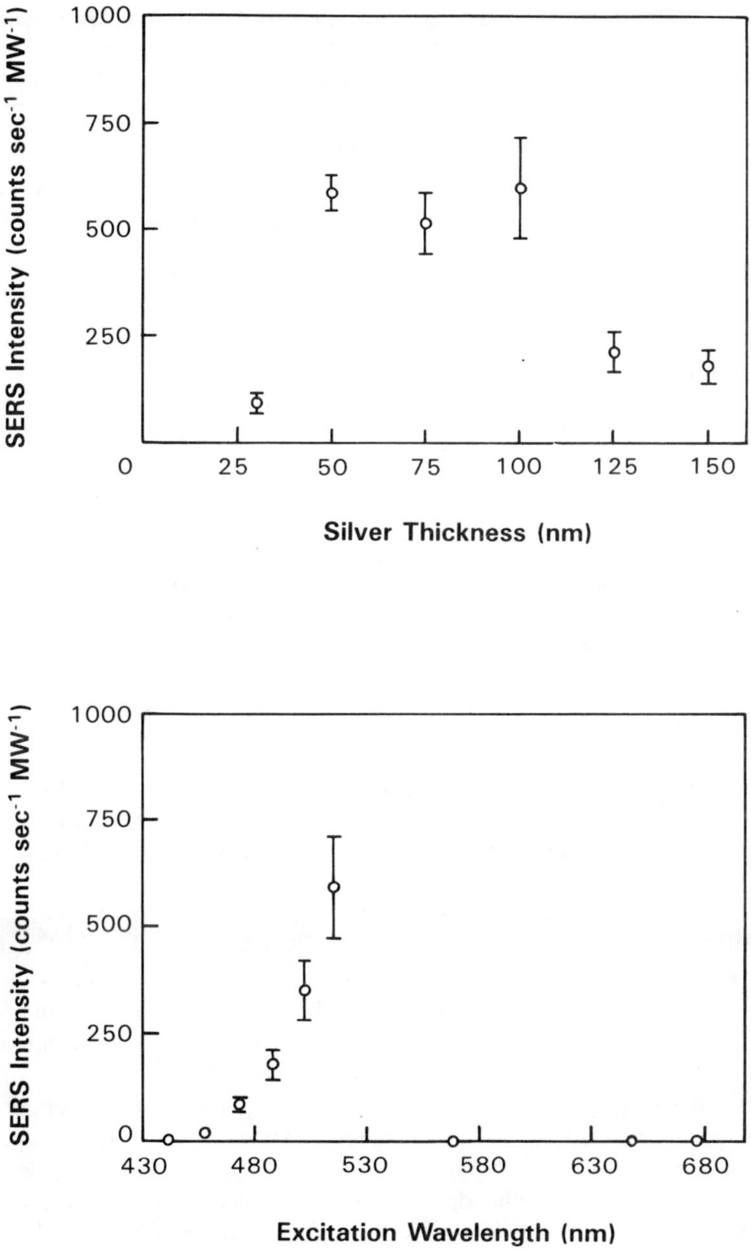

Figure 5.2. Plot of surface enhanced Raman spectroscopy (SERS) intensity (a) versus silver thickness and (b) versus excitation wavelength (11).

Table 5.1. Bulk-phase and surface Raman vibrational frequencies (cm^{-1}) of pyridazine. (Ref. 13, Reprinted with permission from J. Phys. Chem. Copyright 1987, American Chemical Society)

LIQUID	AQUEOUS SOLUTION	SERS −0.55 V	SERS −0.45 V	SERS −0.05 V
363		193	216	224
370	379	381	383	385
622	623	638	641	642
666	667	669	669	670
755				741
		758	765	765
963	975	974	975	982
1063	1068	1066	1066	1070
1116	1123			
		1148		1150
1160	1164	1160	1162	1169
	1184			
		1214	1210	1205
		1234	1224	
1284	1291	1290	1293	1287
				1388
1443	1455	1453	1450	1454
1565				
1570	1573	1575	1575	1576

(two N substitution) and phthalazine (two N substitution in a naphthalene framework) adsorbed onto an Ag electrode has been shown to be due to the donation of nonbonding electrons from the nitrogens to the metal (13). The data for pyridazine are listed in Table 5.1.

The adsorption of 2,2′ -bipyridine onto an Ag electrode in the presence of KCl, KBr, and KI solutions was discussed by Kim and Itoh (14,15). Extension of this research to the adsorption of 2,2′-bipyridine onto Ag colloids with a pH range of 9.0 to 2.7 and in the presence of the anions Cl^-, Br^-, I^-, and SO_4^{2-} was recently described (16). The results of these studies (14–16) were practically identical. That is, the adsorbed species could be described as consisting of four distinct molecular types with specific potentials applied to the electrode and within specific pH ranges. Adsorbed species that were identified included an argentous-type complex (Ag-bipyridine nitrate), a bipyridine adsorbed side-down onto the surface, a strongly bonded species coordinated to Lewis acid sites, and a physisorbed species. These results showed that in some cases good correspondence should exist between species that have been observed on electrode surfaces and those that may be observed on catalytic surfaces.

Surface enhanced spectra (SES) associated with adsorption of ethylene (17,18) and *cis*-2-butene, *trans*-2-butene, isobutylene, or 1-butene (19) onto Ag films have been reported. Raman intensities from the first monomolecular, adsorbed layer were about 200 times larger than intensities from the second layer. In addition, the bonding of the first layer to the surface was achieved through pi-bonds associated with the C=C double bond. The site of such bonding on Ag(110) for adsorbed ethylene has been examined, with the conclusion that adsorption is selective to silver atoms that have a positive charge that can be induced by preadsorption of oxygen (20). Alternately, as proposed by Smoluchowski (21), an electron flow from geometrical irregularities on the surface to the bulk metal would induce a positive charge on the irregularities/surface defects. These possibilities were incorporated into a comprehensive discussion of SERS data from 1-butene and 1,3-butadiene adsorbed onto Ag films (22). The C=C stretching frequency of 1-butene decreased from its 1642 cm^{-1} bulk value to 1605 cm^{-1} upon adsorption. The 1-butene was weakly adsorbed as compared to 1,3-butadiene, and there were at least three distinct types of adsorbed butadiene as compared to one type of adsorbed butene. The C=C stretching frequencies of these three types included 1637 (near the bulk value), 1625, and 1605 cm^{-1}. The species associated with the 1637 cm^{-1} vibrational mode was weakly adsorbed, whereas the other species were tightly bound; such tight bonding was attributed to pi-bonding to positively charged surface ions.

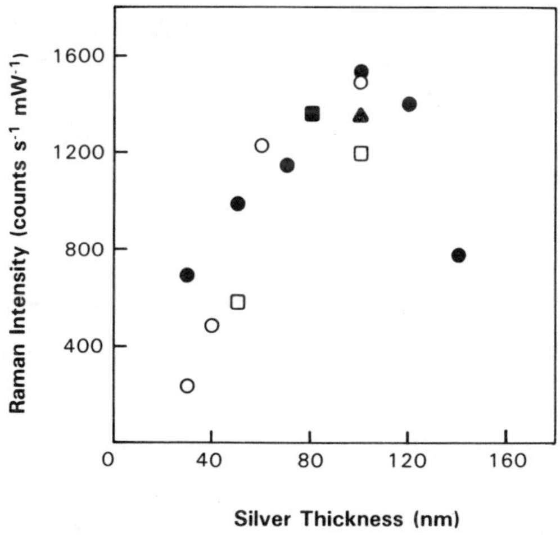

Figure 5.3. Raman enhancement as a function of the silver thickness deposited onto an etched quartz substrate. The intensity of the 1597 cm^{-1} band of adsorbed *p*-nitrobenzoic acid is shown; an excitation wavelength of 514.5 nm was used. The symbols correspond to different angles of evaporation (□: 0°; ○; 40°; ●: 55°; ▲: 65°, and ■: 78° (26). (Reprinted with permission from *The Journal of Physical Chemistry,* Copyright 1985, American Chemical Society.)

The SES of benzene adsorbed onto Ag electrode surfaces in the presence of alcohol solvents was presented by Shin and Kim (23). Instead of the rich spectra usually observed for SERS intensities of pyridine adsorbed on Ag, the benzene ring breathing mode near 1000 cm^{-1} was the primary vibrational mode that experienced significant intensity enhancement. In addition, the benzene ring breathing mode was about an order of magnitude less intense than the corresponding mode for pyridine under identical experimental conditions.

The intensity of the ring breathing mode in *p*-nitrobenzoate as a function of adsorption onto differently prepared Ag surfaces was investigated by Martinez, Gao, and Lopez-Rios (24). Their results showed that the preparation of the surface on which the *p*-nitrobenzoate was adsorbed controlled the observed intensities. A direct correlation between the visible reflectivity of the surfaces and the intensities of SES was presented.

The use of etched quartz for Ag deposition and then *p*-nitrobenzoic acid adsorption, for SERS data acquisition, has been reported (25,26). Such etching produces a rod structure on the surface of the quartz. The dependence of SERS intensities on the thickness of Ag deposited for *p*-nitobenzoic acid adsorption is shown in Figure 5.3. The particle size of the Ag on this support that causes the

Table 5.2. Comparative studies of Raman peak positions for 1-nitropyrene on varous surfaces. (Ref.25, Reprinted with permission from Anal. Chem., Copyright 1986, American Chemical Society)

I. NR PEAK POSITIONS[a] (CM^{-1})	II. SERS PEAK ON SILVER-COATED SPHERES ON GLASS[b] (CM^{-1})	III. SERS PEAK POSITION ON PROLATE POST SURFACE[c] (CM^{-1})
1158	1162	1157
1191	1198	1196
1227	1226	1224
1243	1243	1239
1313	1301	1295
1343	1335	1332
1394	1383	1377
1411		
1426	1423	
1445		
1509	1509	1506
1600	1598	1595
1630	1629	1622

[a]Column I lists the wavenumbers of peaks obtained by a conventional spectrum. [b]Column II lists the wavenumbers of peaks obtained by SERS on 0.364- μm latex spheres coated with 200 nm of silver. [c]Column III lists the wavenumbers of peaks obtained by SERS on prolate posts coated with 80 nm of silver.

greatest enhancement in Raman intensities for benzoquinoline, phenanthrene, coronene, and nitrobenzoic acid was estimated to be between 50 and 100 nm. Interestingly, the data in Figure 5.3 are very close to the data presented by Moody et al. (11) for nitropyrene in Figure 5.2. These data indicate that highly repeatable measurements can be made using the SERS phenomenon. For catalysis, the preparation of samples for such experiments has not been examined thoroughly. However, the above-described results and others discussed below indicate that such preparation should be possible. For example, research on the preparation of supports for SERS and on the practicality of such preparation were published by Enlow et al. (25). Here, glass, paper, and quartz-rod supports were coated with Ag and used to adsorb nitro polynuclear aromatic compounds, including 1-nitropyrene, 9-nitroanthracene, 2-nitronaphthalene, and 2-nitrofluorene. In comparison with NR spectral intensities, the intensities of the SERS spectra from these compounds were enhanced by factors of 5×10^6 to 9×10^6. The limit of detection for 1-nitropyrene was 0.3 ng; a compilation of the bands in 1-nitropyrene is shown in Table 5.2.

Sulfur, Nitrogen, and Carbon Oxides

The use of unsupported Ag powder for observation of SES was demonstrated recently (27,28). The Ag powder, having estimated particle sizes of 1 to 10 μm, was prepared by precipitation from $AgNO_3$ solutions; either air drying or vacuum drying was used to process the powder. Near-application conditions were utilized during adsorption of SO_2 and NO_2 and during spectral acquisition in the presence of O_2 and H_2O. Peak positions for SO_2/Ag were 625, 925, and 960 cm^{-1}; in comparison to the Raman spectra of S-containing Ag compounds (29,30), Ag_2SO_3 (615 and 925 cm^{-1}), and Ag_2SO_4 (968 cm^{-1}), the adsorption of SO_2 was shown to create SO_3^{2-} and SO_4^{2-} surface species. These species were described to be stabilized by the presence of O_2, in a manner similar to the stabilizing effect that adsorbed ligands such as Cl^- have on the SERS of adsorbates in solution for Ag electrodes (31).

The Raman data for NO_2/Ag powder are shown in Figure 5.4. These band positions and their behavior upon evacuation and H_2O exposure (see also Table 5.3) were attributed to the presence of surface NO_2^- and NO_3^- species. The band at 2315 cm^{-1} in Figure 5.4 was associated with adsorbed NO^+ instead of adsorbed N_2 (32). Significant differences in surface chemistry were observed in these experiments when adsorption of NO_2 was monitored by Raman spectroscopy as compared to the case when it was monitored by XPS techniques (28). The equilibrium of NO_2 adsorption, expressed by:

$$2\,NO_2 \leftrightarrows NO_3^- + NO^+$$

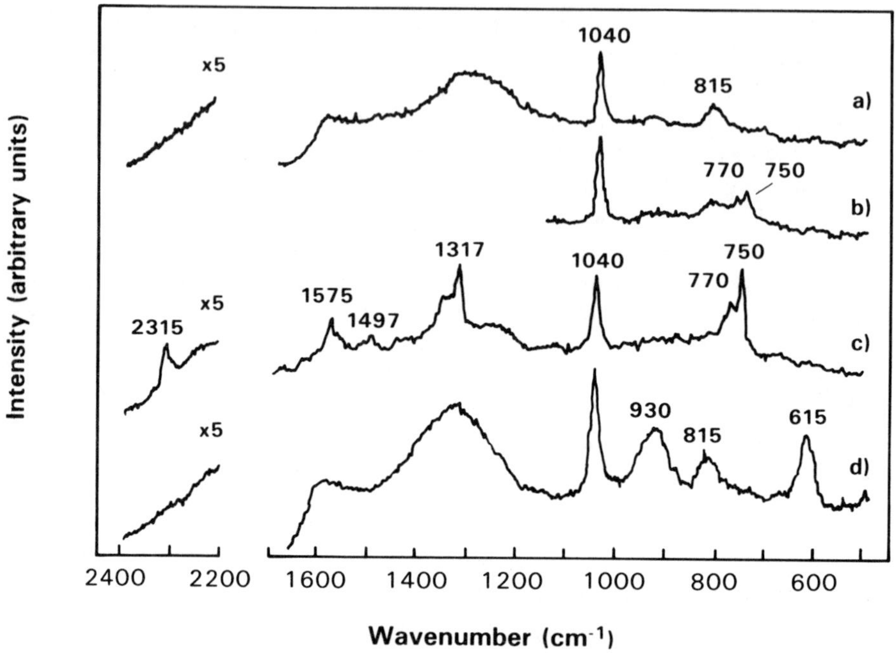

Figure 5.4. Raman spectra of adsorbed nitrogen dioxide on a silver powder surface. (a) Addition of both NO_2 of 10^2 Pa and Ar of 10^5 Pa; (b) evacuation up to 1.3 Pa; (c) evacuation up to 10^{-4} Pa; (d) addition of moistened Ar of $\sim 10^{-5}$ Pa (28).

could be monitored under realistic conditions by Raman experimentation but not by XPS.

The adsorption of O_2, N_2, and CO on porous Ag films was investigated by Pettenkofer et al. (33), who observed SERS enhancements for the Raman bands of CO and N_2, whereas the signal for O_2 was not enhanced so much. If a small amount of O_2 was adsorbed onto the Ag surface before adsorption of other adsorbates such as pyridine and ethylene, enhancements in band intensities of these adsorbates were significantly diminished. These differences were present even though the relative gas phase cross sections of O_2, N_2, and CO are similar—1.2, 1.0, and 0.9, respectively. Hence, in this case the presence of O_2 decreased the concentration of the adsorbed molecules, caused their dissociation, and/or changed surface morphology. For S-containing molecules, as described by Matsuta and Hirokawa (27,28), the presence of O_2 enhanced the stability of adsorbed molecules.

Nanba, Yamamoto, and Ikezawa (34) measured the SERS results for CO adsorbed onto Ag films. Upon exposure of CO to the evaporated Ag/Al sample at 80°K, three distinct Raman bands, at 155, 180, and 2142 cm^{-1}, were observed.

Table 5.3 Raman band positions of Ag-containing nitrogen oxide compounds.

N_2O/Ag	$AgNO_3$	$AgNO_2$
	710	
	730	
815		820
		850
1045	1045	1040
1285	1340	1300
	1600	
	1620	

The 2142 cm^{-1} band was attributed to the C—O vibration of chemisorbed carbon monoxide, whereas the lower frequency modes could not be assigned to a particular surface species. The dependence of the vibrational frequency of C—O on temperature and surface coverage was studied and compared to previous spectroscopic data on the adsorption of CO (35–37). Shifts in the C—O stretching vibration as a function of coverage were shown to be independent of dipole–dipole interactions.

Acrylic Acid

The polymerization of acrylic acid and acrylamide on colloidal silver was studied by SERS (38). Assignments of the observed Raman bands are listed in Table 5.4, along with the assignments of the bands from Na acrylate and liquid acrylic acid. A significant number of the bands for acrylic acid/Ag are at different positions from those observed for Na acrylic. These bands (e.g., 2928, 2858, and 1348 cm^{-1}) were assigned to specific modes by comparison with Raman bands from vinyl polymer (39–41). The bands did not appear immediately after acrylic acid was added to the colloid; instead, a four-minute time lapse occurred before the bands began to grow in intensity. In addition, polarization measurements indicated that at high concentrations (8.6×10^{-3} M) of acrylic acid the ethylene plane of the adsorbed monomer was tilted with respect to the surface. In contrast, at low concentrations (2.1×10^{-4} M) the ethylene plane was oriented parallel to the surface. Intermediate concentrations produced spectra with polarization dependencies indicative of a combination of these configurations. The observation of the polymerization of acrylic acid in this case is a direct consequence of the speed with which Raman spectra could be obtained when such immense enhancements in spectral intensities are present. Comparatively, the

Table 5.4. Assignments of NR and SERS bands for Na acrylate and acrylic acid.

	CM^{-1}		
Na ACRYLATE (REF. 38)	LIQ. ACRYLIC ACID (REF. 38)	ACRYLIC ACID/AG	ASSIGNMENT
3110 (w)	3112	3094	$\nu_{as}(CH_2)$
3043	3039	3034	$\nu_s(CH_2)$
3024	2998	3037	$\nu(CH)$
		2928	$\nu(CH)$
		2858	Polymer
1640		1632	$\nu(C{=}C)$
1550 (w)			$\nu_{as}(CO_2)$
1453			$\nu_s(CO_2)$
1370		1276	$\delta(CH)$
1283			$\nu(C{-}C)$
1053			$\delta(CH_2)$
950 (w)		1348	$\delta(CH_2)$
902		902	$\delta(CO_2)$
836 (w)			$\delta(CH)$
661		665	$\delta(CO_2)$

collection of data by use of NR spectroscopic techniques would be slow as a result of the need to perform either multiple scans and/or noise filtering.

Metal Oxides

The adsorption of tetrahedral d^0 M(VI) oxoanions, CrO_4^{2-}, MoO_4^{2-}, and WO_4^{2-}, onto colloidal Ag was investigated with the use of SERS by Feilchenfeld and Siiman (42). Figure 5.5 shows spectra of adsorbed CrO_4^{2-} for the 600 to 1000 cm^{-1} region as a function of excitation frequency. The maximum enhancement in the Cr—O stretching modes was approximately 1.1×10^5 with an excitation frequency of 580.0 nm. This enhancement was obtained by comparing the 796 cm^{-1} band intensity of chromate adsorbed onto Ag colloid (in reference to the 1020 cm^{-1} band of methanol) with the intensity of the 850 cm^{-1} band of chromate ion in a solution (without Ag) that contained the same amount of methanol as in the reference solution.

The most intense band of MoO_4^{2-} adsorbed onto Ag colloids was located at 865 cm^{-1}; in solution, MoO_4^{2-} has its most intense band at 896 cm^{-1}. For WO_4^{2-} Ag colloids, the SERS data showed that the most intense W—O stretching band was at 895 cm^{-1}, as compared to 930 cm^{-1} when WO_4^{2-} is in solution. Excitation profiles relative to the visible absorption spectra of these tetrahedral anions are shown in Figure 5.6. In each case, a maximum in absorption cor-

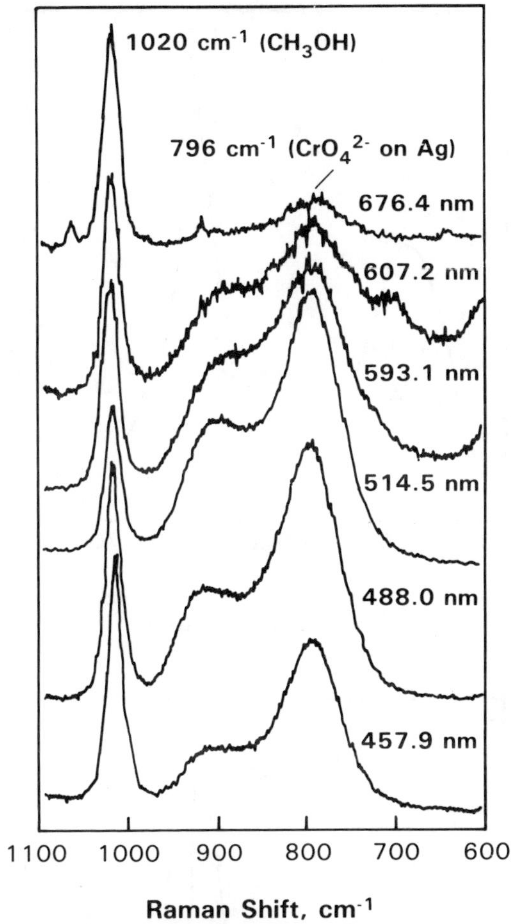

Figure 5.5. Dependence of surface Raman spectra of chromate (3.5×10^{-5} *M*) on colloidal silver in 25 vol% methanol on the excitation wavelength, as indicated in nm (42). Reprinted with permission from *The Journal of Physical Chemistry*. Copyright 1986 American Chemical Society.

responds to a maximum in the excitation profile. Such behavior suggests that the intensities observed in these SERS results were influenced to some degree by resonance Raman enhancements.

The SES of WO_3 on Ag films has been studied (42,43). The WO_3 was prepared by rf sputtering from a tungsten filament in an oxygen plasma. Spectra of these samples showed bands in the 1200 to 1700 cm^{-1} region, whose origins were attributed to carbon; observation of these carbon bands also coincided with the

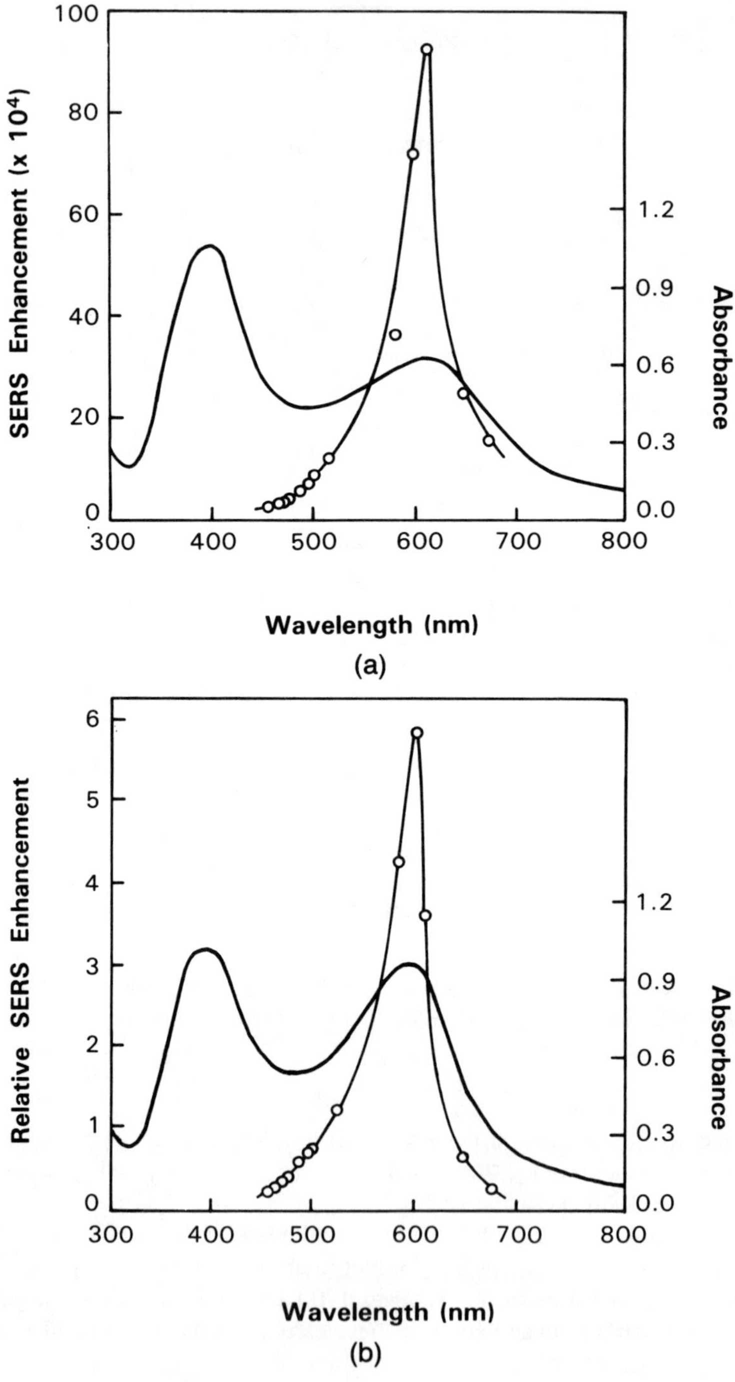
100
80
60
40
20
0
SERS Enhancement (x 10⁴)
1.2
0.9
0.6
0.3
0.0
Absorbance
300
400
500
600
700
800
Wavelength (nm)
(a)
6
5
4
3
2
1
0
Relative SERS Enhancement
1.2
0.9
0.6
0.3
0.0
Absorbance
300
400
500
600
700
800
Wavelength (nm)
(b)

detection of a band at 2930 cm^{-1}. This band was attributed to a C—H vibration, and indicated that the vacuum system in which the samples were prepared contained hydrocarbon impurities. Figure 5.7 shows the SERS results for these samples as a function of the thickness of the Ag film, and Table 5.5 lists assignments of these bands. Maximum enhancement ($\sim 10^3$) in the Raman intensities was observed with an Ag film thickness of 10.0 nm. As can be observed from the data in Figure 5.7 for WO_3/Si, the intensities of the vibrational modes of surface tungsta were greatly reduced when Ag was not present. However, bands at 880, 810, and 350 cm^{-1} for WO_3/Ag are in agreement with those expected for Ag_2WO_4 (44). No SES were obtainable for WO_3 deposited on Pt, Au, or Ni substrates.

Ethylene/Ethylene Oxide

Silver-containing catalysts are unique in their ability to convert ethylene to ethylene oxide, and are quite active in the transformation of methanol to formaldehyde. Hence, McBreen and Moskovits (45) examined two types of Ag-containing catalysts by Raman spectroscopy during the adsorption of ethylene and oxygen. These catalysts included an Ag colloid and Ag electrochemically deposited onto an anodic aluminum oxide. Some of the SES results obtained at elevated temperatures are displayed in Figure 5.8. The intense 1370 and 1596 cm^{-1} bands were attributed to an amorphous surface carbon. Relatively weak peaks at 670 and 985 cm^{-1} experienced a downshift in frequency when $^{18}O_2$ was substituted for $^{16}O_2$; these shifts were in agreement with those expected for isotopic substitution of ^{18}O for ^{16}O if subsurface oxygen were present in the samples. The SERS results were compared with data from EELS (46) and other SERS experiments (47) in which the vibrational frequency of atomic oxygen on Ag was assigned to a 312 cm^{-1} band, whereas frequencies of peroxidic (O_2^{2-}) and superoxidic (O_2^-) species on Ag were assigned to bands at 697 and 1053 cm^{-1}, respectively. Such a large variation in oxygen-associated vibrational modes led to the suggestion that molecular oxygen is present on the surface of silver under conditions employed for epoxidation catalysis.

NH_3

A SERS investigation on NH_3 adsorbed onto Ag electrodes in the presence KBr and KCl was reported by Sanchez, Lombardi, and Birke (48). A listing of the vibrational band positions in relation to those observed in NR spectra of NH_3 is

Figure 5.6. (a) The SERS enhancement profile (o) and absorption spectrum for molybdate (3.75×10^{-3} *M*) on colloidal silver. (b) The SERS excitation profile (o) and absorption spectrum for tungstate (2.6×10^{-3} *M*) on colloidal silver (42).

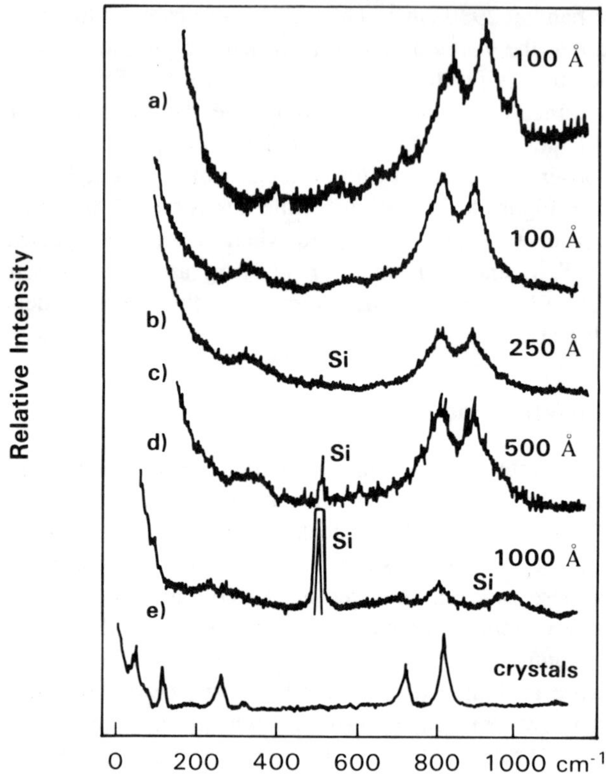

Figure 5.7. Raman spectra of WO_3 thin film in the 100 to 1200 cm^{-1} region: (a) recorded with 488 or 514.5 nm excitation; (b) and (c) WO_3/Ag films with various thickness; (d) WO_3/Si film (100 nm); and (e) WO_3 crystal (43).

given in Table 5.6. Low frequency modes at 216 and 325 cm^{-1} were assigned to Ag—Cl and Ag—N vibrations. The totally symmetric modes of NH_3 were completely depolarized; enhancements in the intensities of the adsorbed NH_3 were estimated to be 3×10^5. Lone pair electron bonding of the NH_3 and the influence of H-bonding were examined by varying the voltage applied to the electrode.

Some other investigations that have employed Ag electrodes as the support for generation of SES of adsorbed molecules include studies of the compound series $C_6H_5(CH_2)_nCOOH$ (49) and propylene carbonate (50). Data from the former study, in which the bands of benzoic acid from NR and SERS data were compared, are listed in Table 5.7. Significant differences in the positions of bands obtained from NR and SERS data can be observed in Table 5.7; the same

Table 5.5 Raman frequencies and relative intensities of WO_3/Ag thin films in the 100 to 1000 cm^{-1} region. (43)

ORTHORHOMIC WO_3^a	WO_3/Si (100 NM)	WO_3/AG/Si 50 NM b)	50 NM c)	10 NM b)	10 NM c)	d)	$Ag_2WO_4^a$	ASSIGNMENT
			1000sh		1000sh			
			935vs		935vs	935m	935s	
		880vs		880vs		880vs	880vs	$\nu(WO_4)$
							860s	
810s[e]	810w	810vs	800vs	810vs	800m	800vs	880w,b	
								$\nu(WO_3)$
720s	715w	720sh				710w		
			680s				700w	
				610vw		610w		
330m		350w		350w		350w	350w	
								$\delta(WO_3)$
276m			300m,b		300w		320w	

[a]Microcrystal (Ventron).
[b]Before thermal treatment.
[c]After thermal treatment during 30 min in air.
[d]Spectrum recorded with λ = 488.1 nm.
[e]Labels sh, vs, w, s, vw, m & b defined in Table 5.6.

observation can be made in comparing bands from the NR and SERS results on propylene carbonate.

Ag Surfaces with Modifying Overlayers

The enhancements in intensities of Raman vibrational modes associated with the adsorbates on Ag surfaces described above also should manifest themselves for other types of molecules associated with Ag surfaces. For example, it may be possible to use Ag substrates as supports on which films or dispersed molecules are placed to enable the acquisition of vibrational spectra of the supported phase. Alternately, decoration of a substrate with Ag may be used to create enhancements in vibrational intensities of otherwise non-SERS active substrates. In this section, recent research will be described that has addressed such scenarios for inducing intensity enhancements.

It has been reported that the addition of metal "adsorbates" to an Ag surface strongly reduces the intensities in SES (51; and see Chapter 2). Gao and Lopez-Rios (52) showed that the presence of thin Si surfaces on Ag decreased intensities of the vibrational modes of pyridine adsorbed onto the Si–Ag surface. However, vibrational spectra of Si layers having an equivalent bulk thickness as small as

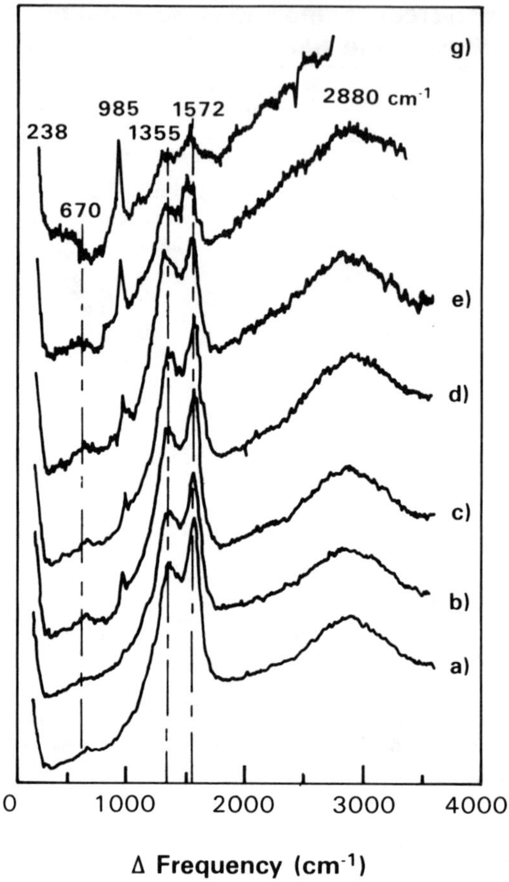

Figure 5.8. The SERS obtained with 488 nm Ar^+ laser excitation for C_2H_4/O_2 (94:6 mixture) adsorbed onto an Ag colloid surface. The sample is reduced and under the gas mixture (a) at 300°K, (b) at 350°K, (c) after 200 minutes at 350°K, (d) at 410°K, (d) at 465°K, (f) subsequently at 455°K, and (g) after 200 minutes at 455°K. The amplification factors used in these spectra were 1, 1.07, 1.6, 1.6, 3.4, 4.0, and 4.21, respectively (45).

0.1 nm have been observed (52), and suggest the possibility of acquiring surface vibrational information previously not obtainable. With further development of overlayer techniques and theoretical considerations, SERS may be used routinely to monitor vibrational spectra of catalytically important coatings under realistic process conditions.

A study of the effect that Pd overlayers on Ag have on the intensities of adsorbed pyridine (53) showed that an equivalent of 0.7 monolayer of Pd would

Table 5.6 Positions (cm^{-1}), intensities, and assignments for Raman bands of ammonia on Ag.[a] (48)

LIQUID	NORMAL RAMAN (14.5 M)	SERS (−0.2 V)	ASSIGNMENT
		216(vs)	νAgCl
		325(s)	νAgN
		675(vw)	δ_r
1046	1126(vw)	1174(w)	δ_d(HNH)
1640	1630(vw)	1624(vw)	δ_d(HNH)
3217	3225(w)		$2\delta_d$
3303	3315(vs)	3248(s)	ν_s(NH)
		3325(vw)	
3387	3405(w)		ν_a(NH)
		3490(b)	ν(OH)

[a]ν = stretch, δ = bend, s (as subscript) = symmetric, d = degenerate, a = antisymmetric, r = rocking, vs = very strong, s (in parentheses) = strong, vw = very weak, w = weak, m = medium, sh = shoulder, b = broad.

decrease by a factor of 10 the intensity of the 1006 cm^{-1} ring breathing mode of pyridine. Subsequent coating of the Pd/Ag surface by an extremely thin layer of Ag restored the 1006 cm^{-1} band intensity to its value before Pd deposition. Such behavior implies that electromagnetic resonances associated with isolated particles do not fully explain SES for pyridine/Ag. A similar situation was found for deposition of Tl and Pb overlayers on Ag (54), whereas the addition of a monolayer equivalent of Cu on Ag shifted pyridine vibrational modes away from their positions on Ag surfaces and produced Raman bands that were not observed for pyridine/Ag (55,56). Other experiments involving competitive adsorption of pyridine, Cl^-, Tl, and SCN^- suggested that only about 3% of the surface of Ag was active for producing SES (57). Such a small percentage of surface atoms that induce enhancements in Raman intensities may preclude the use of Ag, Cu, and Au as substrates on which the deposition of an overlayer metal or metal oxide could be studied under catalytic conditions. However, the influence of Al overlayers on positions and relative intensities of adsorbed pyridine (58) suggests that adsorption on such overlayers can be monitored. In this case, the symmetric, A_1 vibrational modes of pyridine were affected the most by Al overlayers. The intensity of other vibrational modes of pyridine were not decreased to that degree, whereas a new, unidentified mode was activated at 646 cm^{-1}.

The SERS results of the adsorption of oxygen on Al-coated Ag were compared to results of adsorbing oxygen on Ag without the Al overlayer (59). Bands due to peroxidic and superoxidic dioxygen, resulting from different amounts of charge accumulation on antibonding orbitals, were observed. This observation was in agreement with the data of Eickmans, Godmann, and Otto (47). Coverage of the Ag by an equivalent of 0.14 nm of Al decreased the intensity of the Ag—

Table 5.7. The SERS spectra of benzoic acid, NR spectra of benzoate ion and benzoic acid, and their assignment (49).

BENZOIC ACID (NR) ν (cm^{-1})	BENZOATE ION (NR) ν (cm^{-1})	BENZOIC ACID (SERS) ν (cm^{-1})	BENZOIC ACID (SERS) ENHANCEMENT FACTOR[a]	ASSIGNMENT
—	170	—		
—	—	219	s[b]	Ag—O bond
—	404	—		—COO^- rock or ϕ (CC) ν_{16a}
414	—	—		C═COH out-of-plane
—	—	438	s, ‖	COO^- i-p rock or ν_{16a}
616	617	—		α(CCC)ν_6 phenyl ring def.
—	—	676	w	COO^- sym. def.
790	—	—		O═C—OH in-plane bending
—	843	839	s, ‖	ν_{6a}
1001	1001	1001	s, ‖	ν_{12}C—C—C parallel bend
1025	1023	1024		$\nu_{18}\beta$ (HCC)
—	1134	1138	m, ‖	COO^- i-p rock + ν_{6b}
—	1155	—		$\nu_9\beta$ (CH)
1281	—	—		C—OH stretching
1315	1300	—		Ring stretching
—	1392	1384	m, ‖	COO^- sym. stret.
1602	1600	1600	s, ‖	ν_{8a}C—C stret.
1644	—	—		C═O stretching
3072	3074	3065		C—H stret.

[a]Indicates the vibration modes that maintain the molecular plane.
[b]See Table 5.6 for label definition.

O vibrational mode at 330 cm^{-1}, but did not appreciably affect the bands due to O_2^{2-} and O_2^-.

Dorain (60,61) investigated SES of protonation reactions over Ag,Mn mixtures deposited on an Ag electrode. The formation of (Ag,Mn)O on Ag electrodes was shown to occur after oxidation–reduction cycling of electrodes that had been immersed in an alkali halide–$KMnO_4$ solution. This mixed metal oxide was protonated during cathodic voltage swings, producing successively (Ag,Mn)OH and (Ag,Mn)OH_2. Positions of the Raman bands of various protonated mixed metal oxides included 615 cm^{-1} for the (Ag,Mn)—O vibration, 500 cm^{-1} for the (Ag,Mn)—OH vibration, and 415 cm^{-1} for the (Ag,Mn)—OH_2 vibration. The OH stretching vibrations also were analyzed and appeared in the frequency range 3450 to 3520 cm^{-1}. A schematic of the surface species on the (Ag,Mn) surface under oxidizing and reducing conditions is shown in Figure 5.9.

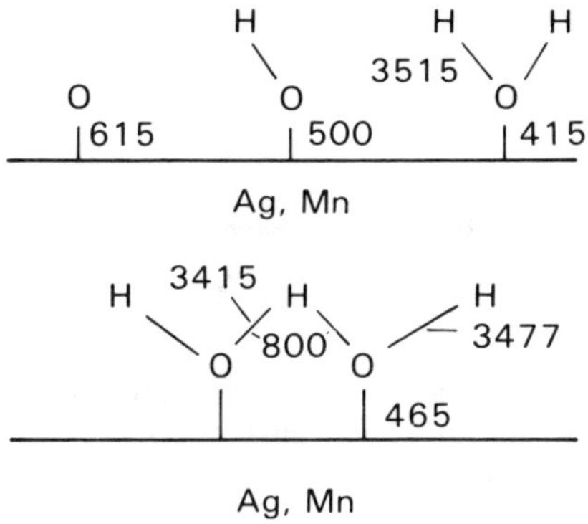

Figure 5.9. Summary of assignments for OH-associated and O-associated species on Ag,Mn substrate (vibrational frequencies are in cm^{-1}) (61). (Reprinted with permission from *The Journal of Physical Chemistry,* Copyright 1986 American Chemical Society.)

The effect of incorporating Pb onto Ag electrodes was examined by Kellogg and Pemberton (62) and Guy, Bergami, and Pemberton (63). In the case of adsorption of HCN on an Ag electrode without Pb (62), the C—N vibrational bands were located at 2130 and 2096 cm^{-1}. The presence of two bands was attributed to two types of bound HCN, an HCN bonded end-on to the surface (2130 cm^{-1}) and HCN side-bound to the surface (2096 cm^{-1}). Incorporation of Pb decreased the intensity of these C—N bands, in a manner similar to the decrease in intensity of pyridine bands upon the incorporation of Pb (63). No changes in the positions of the C—N bands as a function of Pb coverage were described.

Using a Pt substrate, on which extremely small amounts of Ag were deposited, Furtak and Miragliotta (64) and Pemberton (65) investigated pyridine adsorption by use of SERS. This type of experimental surface was designed to elucidate chemical effects of different substrates on characteristic pyridine vibrational bands. No extreme changes in the vibrational modes (other than intensities) were observed. However, this type of experimental design has not been investigated thoroughly for use in monitoring heterogeneous catalysts, where, for example, Ag as an inactive component in a specific catalytic reaction would be incorporated into the catalyst as an initiator of enhanced Raman intensities.

Van Duyne et al. (66) investigated in some detail the effects of thin Ag overlayers on the vibrational spectra of adsorbed transition metal complexes. These results, and others (67,68), suggest that Ag overlayers can produce SES from adsorbates on a wide range of non-SERS substrates. In fact, the type of sites thought to cause enhanced Raman intensities are depicted in Figure 5.10 (66). Of the four sites that are numbered, sites 1 and 2 were considered to be active in the generation of the SES. Such sites are similar to those considered

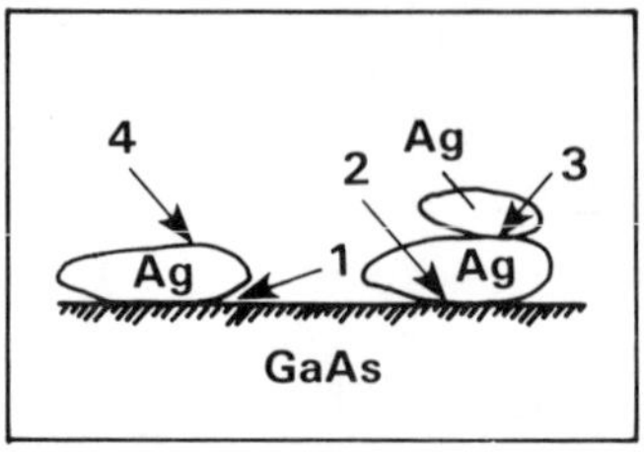

Figure 5.10. Illustration of possible sites causing surface enhanced Raman spectra for Ag-coated GaAs (66). (Reprinted with permission from *The Journal of Physical Chemistry,* Copyright 1985 American Chemical Society.)

to be responsible for activity in catalysts exhibiting strong metal support interactions (SMSI).

A quantitative study of Raman scattering cross sections and enhancement factors for the vibrational modes of pyridine, thiocyanate, isothiocyanate-bridged Cr(III) complexes, Cr(III) and Co(III) hexa-amines, and Os(III) and Os(II) complexes on Ag has been published by Weaver, Farquharson, and Tadayyani (69). These data and the summary of results published by Moskovits (70) provide additional insight into the possible experimental applications of SERS over Ag-containing materials.

Cu SURFACES

Adsorption of N-Containing Species

Copper-containing catalysts are used in the synthesis of aliphatic amines (71). Such amines serve as intermediates in the synthesis of fabric softeners, pigment dispersants, agricultural herbicides and pesticides, drugs, and dyes (72). Possible starting chemicals in these syntheses are alcohols. The reaction scheme for the conversion of such alcohols can be represented by:

$$R'\text{—}OH + NH_2\text{—}R \rightarrow R'\text{—}NH\text{—}R + H_2O$$

Disproportionation and hydrogenolysis (73) and metal nitride or carbide formation (74) can decrease the efficiency of conversion of the alcohol into desired products. Hence, Wokaun et al. (75–77) investigated the SERS of amines and of toluidine, which contains an amino group and aromatic pi-electrons, upon adsorption onto Cu. The samples used in these studies were Cu foils that were SiO_2-blasted and chemically etched. SERS data of *m*-toluidine/Cu are depicted in Figure 5.11. Differences between spectra of the adsorbed surface species and

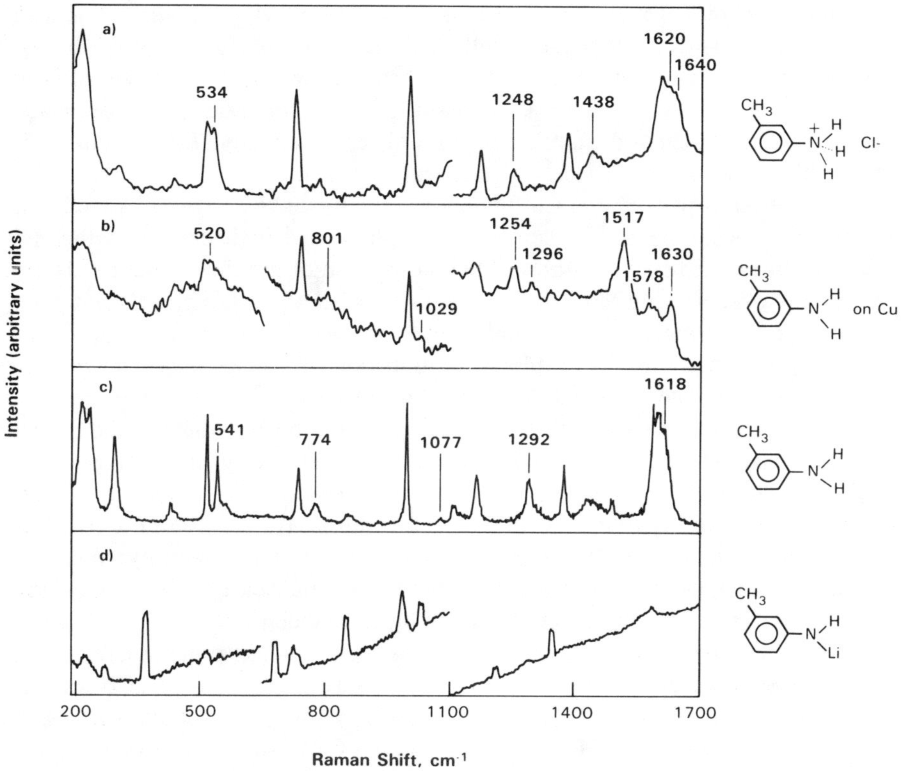

Figure 5.11. Raman spectra of (a) toluidinium chloride, (b) *m*-toluidine adsorbed on a roughened Cu surface, (c) liquid *m*-toluidine, and (d) lithium toluidide (76). (Reprinted with permission from *The Journal of Physical Chemistry,* Copyright 1985 American Chemical Society.)

the neat liquid include an upshifting of the skeletal vibration of the liquid by 12 cm^{-1} to 1630 cm^{-1} when adsorbed on the Cu, a decrease in the aromatic ring vibrational frequencies to 1517 and 1580 cm^{-1} from 1570 and 1605 cm^{-1}, respectively, and the creation of new vibrational bands at 1254 and 801 cm^{-1} upon surface adsorption. A normal coordinate analysis was performed, and the deuterated analogs were studied to show that at room temperature toluidine was bonded via the nitrogen lone pair (R—NH_2 · · · · Cu); spectra were acquired at temperatures as high as 150°C.

The SERS data (75–77) on Cu surfaces also showed that a Cu particle size range of 10 to 100 nm produced the maximum scattering enhancements. Hence, SERS was said to provide a spectroscopic image of the amine interacting with a bulk, metal surface and not with a highly dispersed Cu species.

Creighton (78,79) examined the interaction of N-methylpyridinium cation with Cu colloids and compared the results with those from Ag and Au colloids. Although aggregation of the colloids may have occurred, no indication of such aggregation was discussed. Depolarization ratios and intensities were predicted from theortically derived expressions that utilized an effective Raman tensor for molecules adsorbed at surfaces of spherical particles.

Adsorption of pyridine on Cu fulms was investigated by Albano, Andreasen, and Heras (80). The SERS data were complemented with data from thermal desorption experiments and showed that desorption of pyridine could be clearly identified as containing three distinct states, labeled alpha I, alpha II, and beta. Acquisition of Raman data at temperatures below the maximum of the corresponding desorption peaks enabled discrimination of species important in each state. The beta state involved decomposition of pyridine to unknown products except for creation of a carbon species that coated the Cu. Cracking and partial dehydrogenation of adsorbed pyridine were suggested to occur on these beta sites, which were also suggested to be the sites of catalytic importance. Interestingly, the inability to observe SERS results during the adsorption of pyridine on transition metals was discussed relative to the need for further investigations on these materials, as it is possible that these metals cause the decomposition of pyridine upon its adsorption at near-ambient temperatures.

Adsorption of cyanide on Cu electrodes has been shown to produce a relatively dense layer of an unidentified Cu(I) complex and multilayers of CuCN similar in nature to bulk CuCN (81,82). The importance of oxide surface species in stabilizing adsorbed cyanide was investigated by Mahoney and Cooney (83). Types of CN surface species and their vibrational frequencies and assignments are displayed in Table 5.8. These surface species were shown to be stabilized by a Cu(I) oxide surface layer.

Excitation of SES from molecules adsorbed onto Cu substrates usually requires laser irradiation in the red region of the visible spectrum. Adsorption of colored

Table 5.8. Proposed assignments of ν(CN) subbands.[a] (83)

Cu/CN^- SERS /cm^{-1}	SOLUTIONS $[CN^-]/[Cu]$ /cm^{-1}	ASSIGNMENTS cm^{-1}	ASSIGNMENTS SPECIES
2077	2077	2077	CN^-
2091	2095	2094	$[Cu(CN)_4]^{3-}$
2102	2108	2109.5,2096	$[Cu(CN_3)]^{2-}$
2114	2113	2113	$[Cu(CN)_2]^-$
2130	2137	2137	$[Cu(CN)_2]^-$(?)

[a]Shoulders are partly resolved subbands that vary in intensity with changing potential or concentration.

Table 5.9. SER spectra and band assignments for 2,2′-bipyridine, RBP_4, RP_6, and RBPVP at copper electrodes (all values in cm^{-1}). (84, Reprinted by permission of John Wiley & Sons, Ltd.)

Bpy	RBP_4	RP_6	RBPVP	ASSIGNMENT[a]
1007s	1014w	1005vs		Ring breathing
	1032w		1026w,br	Ring—ring str.
1165m	1170m		1164m	C—C—H def.
	1216w	1213s	1217w	i.p. ring
1269m	1260m		1256m	i.p. ring
1297s				
	1319s		1314s	i.p. ring
1468vs	1478vs	1470w,br	1475vs	i.p. ring, CCH def
	1552s		1547s	i.p. ring
1557m	1557s		1557s	i.p. ring
1593w	1613s	1600s	1604s	i.p. ring

[a]i.p. = in plane; str = stretch; def = deformation.

samples onto Cu surfaces, or on other substrates that potentially will activate enhanced Raman scattering cross sections, can also induce surface enhanced resonant Raman scattering (SERRS). Few such SERRS experiments have been performed on metal substrates. Chambers and Buck (84) published the SERS results for 2,2′-bipyridine, colored Ru-bipyridine, and other pyridine complexes adsorbed onto Cu. These data, which are summarized in Table 5.9, show that SERRS and SERS effects for Cu surfaces are distinct from those observed on Ag substrates (85). The adsorption of the bipyridine complexes was considered to occur through the aromatic rings of the compounds.

Thiourea, thioacetamide, and N,N-dimethylthioform-amide adsorption onto Cu surfaces was examined by the use of SERS (86–88). These compounds, shown in Figure 5.12, were S-bonded to the Cu surfaces instead of N-bonded.

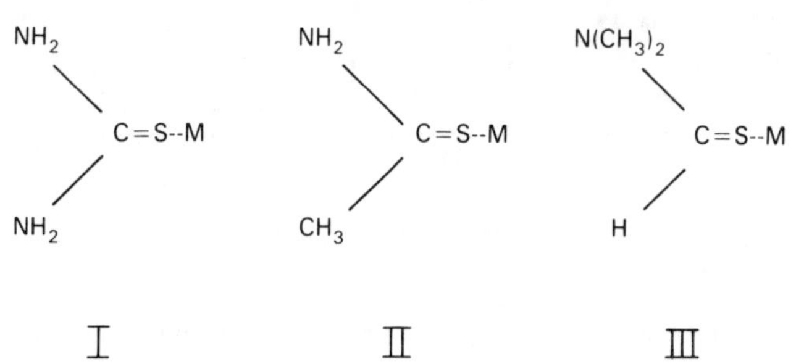

Figure 5.12. Adsorption geometry of thiourea (I), thioacetamide (II), and N,N,-dimethylthioformamide (III) on metal (M) surfaces (88).

The C═S vibrational frequencies were downshifted subsequent to adsorption, indicative of a possible decrease in the C═S bond order.

Enhancements in the Raman scattering cross sections of molecules adsorbed onto Cu (and Ag) can be increased by suitable preparation of their surfaces. For example, Thierry and Leygraf (89) showed that microlithographically prepared Cu surfaces were relatively more amenable to repeatably producing SERS as compared to other sample preparation methods. Additionally, they showed that 1,2,4-triazole adsorbed on Cu in a flat configuration (89).

Au SURFACES

Adsorption of Organic Molecules

Shifts in the vibrational frequencies of alkenes upon adsorption onto transition metal surfaces are a result of pi–*d* electron interactions where donation of electron density from the pi bond of the alkene to *d* orbitals of the metal occurs, in addition to back-donation of electron density from the *d* orbitals of the metal to empty pi* antibonding orbitals of the alkene. Both of these interactions decrease the C═C bond order with a concomitant decrease in the C═C symmetric stretch-

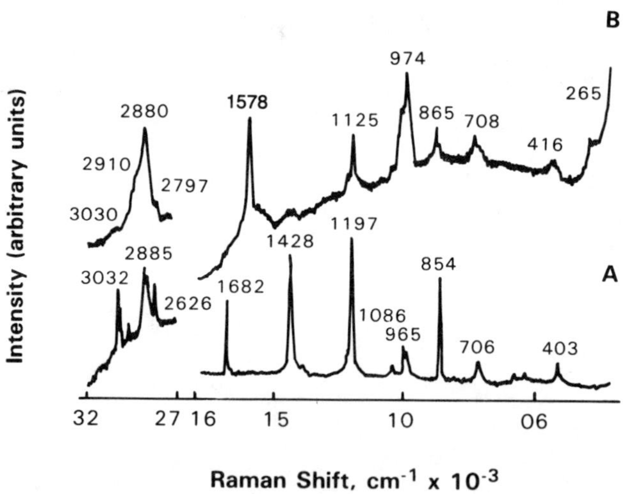

Figure 5.13. Comparison of normal Raman (NR) and surface enhanced Raman (SER) spectra of 1,4-cyclohexadiene with (a) NR spectrum of a liquid, laser power 50 mW at 514.5 nm excitation, and (b) SER spectrum of adsorbed species, laser power 30 mW at 647.1 nm excitation (92). (Reprinted with permission from *The Journal of Physical Chemistry*, Copyright 1985 American Chemical Society.)

Table 5.10. Bulk-phase and surface Raman vibrational frequencies (cm^{-1}) and assignments for 1,4-cyclohexadiene (92).

NORMAL RAMAN[a]	SER[b]	VIBRATIONAL ASSIGNMENT
403 m	400 m	Ring pucker
	416 m	
530 w		Ring deformation
574 w		Ring deformation
706 m	708 m	Out-of-plane HC═CH bend
854 s	853 m	Ring breathing
985 m	974 s	Out-of-plane HC═CH bend
1035 w	1030 w	In-plane HC═CH bend
1197 s	1175 m	In-plane HC═CH bend
1377 w		Ring stretch
1428 s	1425 w,br	CH_2 scissors
1682 s	1578 s	Ring C═C stretch
2826 m	2797 w	CH_2 stretch
2864 m	2850 s	CH_2 stretch
2885 m	2910 m	CH_2 stretch
2975 m	2940 m	CH_2 stretch
3032 m	3030 w	═CH stretch
3073 w		═CH stretch

[a]Obtained for liquid 1,4-cyclohexadiene. [b]Surface enhanced Raman spectra for 1,4-cyclohexadiene at gold, 0 mV vs. SCE.

ing frequency. It has been postulated that bonding in bulk-phase alkene–transition metal complexes is similar to that described above (90). Heuristically, in the case of adsorption onto Au, having an s^1d^{10} configuration, donation of electron density from the alkene pi orbital to the half-filled *s* orbital also can cause alkene bonding. Consequences of such bonding for ethylene should include decreased C═C stretching and CH_2 bending frequencies; SERS has been used to monitor such decreased vibrational frequencies (91).

The SERS results for 1-pentene, 1-hexene, cyclopentene, 1,4-cyclohexadiene, 1,3-cyclohexadiene, phenylacetylene, 2-butyne, and 1-pentyne adsorbed onto Au electrodes have been described (92). A comparison of the NR and SERS data for 1,4-cyclohexadiene is presented in Figure 5.13 and Table 5.10. The C═C stretching frequency decreased from 1682 cm^{-1} in the NR data to 1578 cm^{-1} in the SERS data. This decrease in vibrational frequency was attributed to pi-bonding to the Au surface. For ethylene and 1,4-cyclohexadiene, the in-plane and out-of-plane HC═CH bending mode frequencies also decreased upon adsorption (93). The most intense modes in the NR and SERS results were associated with the C═C moiety; relative intensities in the NR data were different from those in the SERS data. On the other hand, the CH_2 scissors vibration at 1428 cm^{-1} was the most intense band in the NR data but was extremely weak in the SERS data. Carbon–hydrogen stretching vibrational frequencies in the SERS results followed a trend indicative of hybridization of the adsorbed species; such hybridization was proposed by Ibach, Hopster, and Sexton (94) in their investigations using EELS experimentation. In addition, surface selection rules were proposed (93) to be more important for SERS spectra as compared to NR spectra. These selection rules helped to describe the orientation of the adsorbed olefin with respect to the Au surface.

Observation of vibrational modes of acetylene adsorbed onto Au surfaces by SERS methods was not successful (92). It was concluded that the high reactivity of acetylene produced polyacetylene immediately upon adsorption (95,96). This conclusion was in agreement with the high reactivity of acetylene upon interaction with Rh surfaces (95). Phenylacetylene, however, was observed to produce a very rich SERS spectrum when adsorbed onto Au substrates. The geometry of phenylacetylene on Au surfaces was determined to be with the triple C–C bond parallel to the surface, a result in agreement with results for phenylacetylene on Cu and Ag electrodes and with the assignment of its vibrational modes from NR spectra (97,98).

Partial oxidation of ethylene subsequent to its interaction with Au substrates was investigated by Patterson and Weaver (99). The major products formed during electrooxidation over Au electrodes at low pH were acetaldehyde and ketones (100). The SERS data were consistent with the formation of a $CH_3CH_2O^{2-}$ intermediate, identified by intense bands at 990 and 425 cm^{-1}.

Adsorption of Oxides

Gold does not form an oxide surface layer to the extent that an oxide forms on Ag and Cu. As a consequence, acid–base reactions should not occur over Au upon exposure to gaseous acidic oxides such as NO or NO_2. Hence, no SERS spectra of NO_3^-/Au were observed (101). An identical situation should occur for the interaction of SO_2 and CO_2 with finely divided Au, whereas on Ag surfaces strong SO_2 and CO_2 SERS data have been observed. Recent studies involving oxidation–reduction and complexation reactions of cyanides on Au colloid surfaces have produced interesting SERS results (101). For example, the major peak observed was the symmetric C—N stretching mode, the frequency of which shifted significantly as a result of coordination changes with the Au (see Table 5.11). The formation of $Au(CN)_3^{2-}$ was postulated when excess CN^- was present in solution; oxidation of an adsorbed $Au(CN)_2^-$ species also produced the square-planar, $Au(CN)_4^-$ adsorbed species. In an electrode environment using Na_2SO_4 aqueous solutions, the C—N vibrational frequency of adsorbed thiocyanide was located in the 2110 to 2130 cm^{-1} region (102). A complementary infrared investigation on this Au electrode system showed that the vibrational frequencies in the IR data were very close to those observed in the SERS data.

Table 5.11. Symmetric mode frequencies for Au cyanide compounds (101)

MOLECULE	$\nu_1[a_{1g}$ (cm^{-1})
AuCN	2261
$Au(CN)_2$	2147
$Au(CN)_4$	2209
$Au(CN)_2Cl_2$	2199
$Au(CN)_2Br_2$	2192
$Au(CN)_2I_2$	2186

The SES of CO adsorbed onto an Au electrode surface in the presence of $HClO_4$ solutions was reported by Tadayyani and Weaver (103). The C—O stretching frequency varied between 2080 and 2110 cm^{-1}, depending on the voltage at the electrode. At the point of zero charge this vibrational frequency was at 2105 cm^{-1}, in excellent agreement with vibrational data for CO adsorbed on catalyst surfaces.

Surface Overlayers and Colloids

As discussed in the section on "Ag Surfaces" (above), the covering of an Ag electrode surface with a small amount of SERS inactive metal usually causes a substantial decrease in the Raman vibrational intensities of most adsorbed species. For Au surfaces, thin layers of Hg, Tl, and Pb also decreased SERS intensities (104). However, for adsorption of M—X (X = Cl or Br) on such surfaces, the band intensities of the M—X vibrational mode were as intense after modification as before modification of the Au surface with a layer of otherwise non-SERS active metal. The M—X vibrational positions, near 250 cm^{-1} for Hg—Cl and 190 cm^{-1} for Hg—Br compounds, were downshifted on application of the various coatings onto Au as compared to their bulk values; such downshifting is indicative of smaller halide–surface bond orders. Similar results were obtained for the adsorption of thiocyanate and pyridine. These results, as compared to similar SERS results for Ag substrates, were interpreted in terms of enhanced stability of SERS-required surface roughness on Au when the surface was modified by application of an overlayer metal (105). Such information is invaluable in the design of experimental procedures for elucidation of surface interactions on metals other than Au, Ag, and Cu.

High pressure SERS studies of adsorption complexes formed upon interaction of pyridine, tetracyanaquinodimethane (TCNQ), tetrathiafulvalen (TTF), and 4-poly(vinyl)pyridine with Au colloid surfaces were presented by Sandroff, King, and Hershbach (106). These studies were facilitated by the design and use of a diamond-anvil cell that allowed spectra to be obtained at pressures as large as 26 kbar. Dependence of the ring breathing mode of pyridine upon increasing the pressure was found to be 0.3 cm^{-1}/kbar. Band broadening behavior of this mode with increasing pressure indicated that the pyridine was adsorbed to the colloid via the nitrogen lone pair electrons. These data extended previously published interpretations (107) about the surface adsorption behavior of pyridine under high pressures. In the case of TCNQ and TTF adsorbed on the Au colloid, the species that were identified included $TCNQ^-$ and TTF^+. The TFF^+ was attached to acceptor sites and the $TCNQ^-$ to donor sites. The success of this high pressure investigation suggests that similar techniques should be used in future experiments to elucidate reaction rates and to identify adsorbed species.

OTHER SERS DATA

Two alkalis (Na, K) commonly used as promoters in catalytic reactions have been reported to induce SES (108). The interaction of NH_3 with Na clusters, of approximately 10.0 nm diameter, was investigated by Schulze et al. (109). A comparison of these data with other data on adsorbed NH_3 is presented in Table 5.12. Continued laser irradiation of the NH_3/Na sample caused decomposition of NH_3 to N_2 and H_2; this decomposition was accompanied by the formation of SERS bands indicative of a Na—NH_2 species.

Ammonia synthesis catalysts commonly contain K as a promoter for improving the yield of ammonia. Hence, interaction of K with N_2, as monitored by SERS, was investigated (110). The diameter of the K particles was estimated to be near

Table 5.12. Comparison of experimental results for ammonia and its fragments (109).

		VIBRATIONAL MODE			
	M-N	1	2	3	4
NH_3 gas	—	3336/3337	932/968	3414	1627.5
NH_3 solid	—	3223	1060	3378	1646
NH_3/Pt[a]	—	3240	1140	3340	1600
NH_3/Pt[b]	350	3150	1190	3320	1630
NH_3/Na_x	380	3205	1138	3370	1625
NH_2/matrix			1499	3220	
NH_2/surface		3290	1610	3380	
NH_2/Na_x	550	3280	1540	3380	
NH/matrix-Na_x		3133	3220		
N_2/Na_x		2325			
H_2/Na_x		4425			

[a]Low coverage.
[b]High coverage.

Table 5.13. Band positions and FWHMs of the vibrational bands for N_2/K (110).

$^{14}N_2$		$^{15}N_2$	
(CM^{-1})	FWHM (CM^{-1})	(CM^{-1})	FWHM (CM^{-1})
273	60	270	60
1822	31	1760	30
2286	37	2213	38
2328	3	2251	≤3
3611	43	3486	46

10.0 nm; N_2 adsorption was facilitated by matrix isolation techniques with the sample temperature maintained at 10°K. Isotopic substitution of $^{14}N_2$ by $^{15}N_2$ was used to confirm spectral interpretation and produced the positions of bands that are listed in Table 5.13. The presence of N_2^- was suggested by observation of bands located at 1822 and 1760 cm^{-1} (111). Analysis of the N_2/Na system by SERS failed to detect Raman bands that could be attributed to a reduced N_2 surface species.

Parker, Siedler, and Hexter (112) have described SERS results for Pd/Al_2O_3 and Pd/SiO_2 hydrogenation catalysts. The SERS intensities of adsorbed C_2D_2 were dependent on the presence of H_2 and/or D_2 in the gaseous phase over the catalyst. The Pd on the Al_2O_3 support produced a dramatically increased scattering intensity as compared to Pd on the SiO_2 support.

REFERENCES

1. Fleischmann, M., P. J. Hendra, and A. J. McQuillan, *Chem. Phys. Lett. 26,* 123 (1974).
2. Jeanmaire, D. L., and R. P. Van Duyne, *J. Electroanal. Chem. 84,* 1 (1977).
3. Albrecht, M. G., and J. A. Creighton, *J. Am. Chem. Soc. 99,* 5215 (1977).
4. Seki, H., *J. Electron Spectrosc. and Related Phenom. 39,* 289 (1986).
5. Murray, C. A., in *Advances in Laser Spectroscopy,* vol. 3, ed. B. A. Garetz and J. R. Lombardi (Wiley, New York, 1986), p. 49.
6. Furtak, T. E., and D. Roy, *Phys. Rev. Lett. 50,* 1301 (1983).
7. Pettinger, B., and L. Moerl, in *Studies in Surface Science and Catalysis,* vol. 14, ed. C. R. Brundle and H. Morowitz (Elsevier, Amsterdam, 1983).
8. Chang, R. K., and T. E. Furtak, eds., *Surface Enhanced Raman Scattering* (Plenum, New York, 1982).
9. Cardona, M., and G. Guntherodt, eds., *Light Scattering in Solids,* vol. IV (Springer, New York, 1983).
10. Wohaun, A., Solid State Physics, vol 38, Ehrenreich, H., F. Seitz, and D. Turnbull, eds., (Academic Press, New York, 1984).
11. Moody, R. L., T. Vo-Dinh, and W. H. Fletcher, *Appl. Spectrosc. 41,* 966 (1987).
12. Ikezawa, Y., H. Saito, K. Matoui, and G. Toda, *Surf. Sci. 176,* 603 (1986).
13. Takahashi, M., M. Niwa, and M. Ito, *J. Phys. Chem. 91,* 11 (1987).
14. Kim, M., and K. Itoh, *J. Electroanal. Chem. 188,* 137 (1985).

15. Kim, M., and K. Itoh, *Chem. Lett.*, 357 (1984).
16. Kim, M., and K. Itoh, *J. Phys. Chem. 91,* 126 (1987).
17. Pockrand, I., *Chem. Phys. Lett. 92,* 514 (1982).
18. Pockrand, I., *Surf. Sci. 126,* 192 (1983).
19. Moskovits, M., and D. P. DiLella, *Chem. Phys. Lett. 73,* 500 (1980).
20. Backx, C., C. P. M. de Groot, and P. Biloen, *Appl. Surf. Sci. 6,* 256 (1980).
21. Smoluchowski, R., *Phys. Rev. 60,* 661 (1941).
22. Itoh, K., M. Tsukada, T. Koyama, and Y. Kobayashi, *J. Phys. Chem. 90,* 5286 (1986).
23. Shin, G.-Seung, and J.-Jean Kim, *Chem. Phys. Lett. 120,* 569 (1985).
24. Martinez, J. L., Y. Gao, and T. Lopez-Rios, *Phys. Rev. B 33,* 5917 (1986).
25. Enlow, P. D., M. Buncick, R. J. Warmack, and T. Vo-Dinh, *Anal. Chem. 58,* 1119 (1986).
26. Meier, M., A. Wokaun, and T. Vo-Dinh, *J. Phys. Chem. 89,* 1843 (1985).
27. Matsuta, H., and K. Hirokawa, *Appl. Surf. Sci. 27,* 482 (1987).
28. Matsuta, H., and K. Hirokawa, *Surf. Sci. 172,* L555 (1986).
29. Dorain, P. B., K. U. Von Raben, R. K. Chang, and B. L. Laube, *Phys. Lett. 84,* 405 (1981).
30. Chen, T. T., K. U. Von Raben, D. V. Murphy, R. K. Chang, and B. L. Laube, *Surf. Sci. 143,* 369 (1984).
31. Von Raben, K. U., P. B. Dorain, T. T. Chen, and R. K. Chang, *Chem Phys. Lett. 95,* 269 (1983).
32. Cooney, R. P., and P. Tsai, *J. Raman. Spectrosc. 9,* 33 (1980).
33. Pettenkofer, C., J. Eickmans, U. Erturk, and A. Otto, *Surf. Sci. 151,* 9 (1985).
34. Nanba, T., I. Yamamoto, and M. Ikezawa, *J. Phys. Soc. (Japan) 55,* 2716 (1986).
35. DiLella, D. P., A. Gohin, R. H. Lipson, P. McBreen, and M. Moskovits, *J. Chem. Phys. 73,* 4282 (1980).
36. Abe, H., K. Manzel, W. Schluze, M. Moskovits, and D. P. Di Lella, *J. Chem. Phys. 74,* 792 (1981).
37. Seki, H., *Solid State Commun. 42,* 695 (1982).
38. Suh, J. S., and K. H. Michaelian, *J. Phys. Chem. 91,* 598 (1987).
39. Feairheller, W. R., and J. E. Keaton, *Spectrochim. Acta A 23,* 2225 (1967).
40. Bardet, L., G. Cassanas-Fabre, and M. Alain, *J. Mol. Struct. 24,* 153 (1975).
41. Gupta, M., and R. Bansil, *J. Polym. Sci. Polym. Phys. Ed. 19,* 353 (1981).
42. Feilchenfeld, H., and O. Siiman, *J. Phys. Chem. 90,* 2163 (1986).
43. Pham Thi, M., *Chem. Phys. Lett. 115,* 130 (1985).
44. Turkovic, A., D. L. Fox, J. F. Scott, S. Geller, and G. F. Ruse, *Mat. Res. Bull. 52,* 189 (1977).
45. McBreen, P. H., and M. Moskovits, *J. Catal. 103,* 188 (1987).
46. Backx, C., C. P. M. De Groot, P. Biloen, and W. H. M. Sachtler, *Surf. Sci. 128,* 81 (1983).
47. Eickmans, J., A. Godmann, and A. Otto, *Surf. Sci. 127,* 153 (1983).
48. Sanchez, L. A., J. R. Lombardi, and R. L. Birke, *Chem. Phys. Lett. 108,* 45 (1984).
49. Chaozhi, W., H. Tianjing, G. Xiaoping, L. Jungqing, X. Houwen, and L. Fan-Chen, *J. Mol. Struct. 140,* 227 (1986).
50. Hill, I. R., D. E. Irish, and G. F. Atkinson, *Langmuir 2,* 752 (1986).
51. Gao, Y., and T. Lopez-Rios, *Phys. Rev. Lett. 53,* 2583 (1984).
52. Gao, Y., and T. Lopez-Rios, *Solid State Commun. 60,* 55 (1986).
53. Lopez-Rios, T., and Y. Gao, *J. Vac. Sci. Technol. B 3,* 1539 (1985).
54. Watanabe, T., N. Yanagihara, K. Honda, B. Pittinger, and L. Moerl, *Chem. Phys. Lett. 96,* 649 (1983).
55. Moerl, L., and B. Pettinger, *Solid State Commun. 43,* 315 (1982).
56. Pettinger, B., and L. Moerl, *J. Phys. (Paris) 44,* C10 (1983).
57. Furtak, T. E., and D. Roy, *Phys. Rev. Lett. 50,* 1301 (1983).

58. Lopez-Rios, T., Y. Gao, and G. Vuye, *Chem. Phys. Lett. 111,* 249 (1984).
59. Gao, Y., G. Vuye, and T. Lopez-Rios, *Surf. Sci. 164,* L819 (1985).
60. Dorain, P. B., *J. Phys. Chem. 90,* 5808 (1986).
61. Dorain, P. B., *J. Phys. Chem. 90,* 5812 (1986).
62. Kellogg, D. S., and J. E. Pemberton, *J. Phys. Chem. 91,* 1126 (1987).
63. Guy, A. L., B. Bergami, and J. E. Pemberton, *Surf. Sci. 150,* 226 (1985).
64. Furtak, T. E., and J. Miragliotta, *Surf. Sci. 167,* 381 (1986).
65. Pemberton, J. E., *J. Electroanal. Chem. 167,* 317 (1984).
66. Van Duyne, R. P., J. P. Haushalter, M. Janik-Czachor, and N. Levinger, *J. Phys. Chem. 89,* 4055 (1985).
67. Murray, C., and D. Allara, *J. Chem. Phys. 76,* 1290 (1982).
68. Tsang, J. C., Ph. Avouris, and J. R. Kirtley, *J. Chem. Phys. 79,* 493 (1983).
69. Weaver, M. J., S. Farquharson, and M. A. Tadayyoni, *J. Chem. Phys. 82,* 4867 (1985).
70. Moskovits, M., *Rev. Mod. Phys. 57,* 783 (1985).
71. Baiker, A., W. Caprez, and W. Holstein, *Ind. Eng. Chem. Prod. Res. Dev. 22,* 323 (1983).
72. Baiker, A., and J. Kijenski, *Catal. Rev. Sci. Eng. 27,* 653 (1985).
73. Kijenski, J., J. Burger, and A. Baiker, *Appl. Catal. 11,* 295 (1984).
74. Baiker, A., D. Monti, and Y. Song Fan, *J. Catal. 88,* 81 (1984).
75. Wokaun, A., A. Baiker, W. Fluhr, M. Meier, and S. K. Miller, *J. Vac. Sci. Technol. B 3,* 1397 (1985).
76. Wokaun, A., A. Baiker, S. K. Miller, and W. Fluhr, *J. Phys. Chem. 89,* 1910 (1985).
77. Wokaun, A., and A. Baiker, *Forsch. Chemia 40,* 2 (1986).
78. Creighton, J. A., *Surf. Sci. 158,* 211 (1985).
79. Creighton, J. A., M. S. Alvarez, D. A. Weitz, S. Garoff, and M. W. Kim, *J. Phys. Chem. 87,* 4793 (1983).
80. Abano, E., G. Andreasen, and J. Heras, *Chem. Phys. Lett. 124,* 567 (1986).
81. Laufer, G., T. F. Schaaf, and J. T. Huneke, *J. Chem. Phys. 73,* 2973 (1980).
82. Benner, R. E., K. U. Von Raben, R. Dornhaus, and R. K. Chang, *Surf. Sci. 102,* 7 (1981).
83. Mahoney, M. R., and R. P. Cooney, *J. Chem. Soc. Faraday Trans. I 81,* 2123 (1985).
84. Chambers, J. A., and R. P. Buck, *J. Raman Spectrosc. 15,* 160 (1984).
85. Blatchford, C. G., J. R. Campbell, and J. A. Creighton, *Surf. Sci. 108,* 411 (1981).
86. Loo, B. H., *Chem. Phys. Lett. 89,* 346 (1982).
87. Loo, B. H., and Y. G. Lee, *Chem Phys. Lett. 112,* 580 (1984).
88. Loo, B. H., Y. G. Lee, and Z Yazid, *Chem. Phys. Lett. 114,* 405 (1985).
89. Thierry, D., and C. Leygraf, *J. Electrochem. Soc. 133,* 2236 (1986).
90. Felter, T. E., and W. H. Weinberg, *Surf. Sci. 103,* 265 (1981).
91. Patterson, M. L., and M. J. Weaver, *J. Phys. Chem. 89,* 1331 (1985).
92. Patterson, M. L., and M. J. Weaver, *J. Phys. Chem. 89,* 5046 (1985).
93. Stidman, H. D., *Spectrochim. Acta 21,* 23 (1965).
94. Ibach, H., H. Hopster, and B. Sexton, *Appl. Surf. Sci. 1,* 1 (1977).
95. Parker, W. L., A. R. Siedle, and R. M. Hexter, *J. Am. Chem. Soc. 107,* 264 (1985).
96. Harada, I., Y. Furukawa, M. Tasumi, H. Shirakawa, and S. Ikeda, *J. Chem. Phys. 73,* 4746 (1980).
97. Arbantes, L. M., M. Fleischman, I. R. Hill, L. M. Peter, M. Mengoli, and G. Zotti, *J. Electroanal. Chem. 164,* 177 (1984).
98. Wotiz, J. H., and F. A. Miller, *J. Am. Chem. Soc. 71,* 3441 (1949).
99. Patterson, M. L., and M. J. Weaver, *J. Phys. Chem. 89,* 1331 (1985).
100. Herlem, M., F. Bobilliart, and A. Thiebault, in *Encyclopedia of Electrochemistry of the Elements,* ed. J. A. Bard and H. Lund, Marcel Dekker, New York, vol. XI (1978).
101. Dorain, P. B., and K. U. Von Raben, *Surf. Sci. 160,* 164 (1985).

102. Corrigan, D. S., J. K. Foley, P. Gao, S. Pons, and M. J. Weaver, Off. Naval Res. Tech. Report #46, N00014-79-C-0670 (1985).
103. Tadayyani, M. A., and M. J. Weaver, *Langmuir 2,* 179 (1986).
104. Leung, L.-W. H., and M. J. Weaver, *J. Electroanal. Chem. 217,* 367 (1987).
105. Fleischmann, M., and Z. Q. Tian, *J. Electroanal. Chem. 217,* 385 (1987).
106. Sandroff, C. J., H. E. King, Jr., and D. R. Hershbach, *J. Phys. Chem. 88,* 5647 (1984).
107. Podini, P., and J. M. Schnur, *Chem. Phys. Lett. 93,* 86 (1982).
108. Schulze, W., B. Breithaupt, K.-P. Charle, and U. Kloss, *Ber. Bunsenges. Physik Chem. 88,* 308 (1984).
109. Schulze, W., B. Breithaupt, F. Frank, and F. W. Froben, *Surf. Sci, 156,* 963 (1985).
110. Schulze, W., K.-P. Charle, and U. Kloss, *Surf. Sci. 156,* 822 (1985).
111. Holzer, W., W. F. Murphy, and H. J. Bernstein, *J. Mol. Spectrosc. 32,* 13 (1969).
112. Parker, W. L., A. R. Siedle, and R. M. Hexter, *J. Catal. 99,* 482 (1986).

6
Raman Spectroscopy of Molecules in Solution

INTRODUCTION

Although more infrared than Raman spectroscopic information has been published on molecules in solution, Raman spectroscopic analysis of solutions has provided important information, which by itself or in concert with infrared analyses has been useful in investigators' understanding of catalyst preparation and use. Identification of the type and symmetry of species present as a function of solution properties, such as pH, type of solvent, solute concentrations, temperature, and so on, can be accomplished by application of Raman techniques. These solution species, whether used in homogeneous catalytic reactions or as precursors for the synthesis of heterogeneous catalysts, can control the activity and selectivity of catalytic reactions. Hence, this chapter will summarize spectroscopic investigations of molecular and ionic species in solution.

First, a section on "Normal Raman Investigations of Solutions" summarizes some normal Raman (NR) spectroscopic data on small organic and inorganic molecules. This summary is meant to give background information on identifying molecular symmetries and species in solution and to present recent examples of such studies. This information serves as a starting point for interpretation of Raman spectra and has been used repeatedly in defining the type and relative concentration of supported catalytic phases (see Chapter 4).

A section on "Resonance Raman Analysis of Solutions" then summarizes recent investigations that have employed resonance Raman techniques to identify molecular and ionic species in solution. Here, information is presented about the substrates on which molecules are adsorbed, and there is detailed discussion of vibrational activity in solution and in surface-supported species. Molecular symmetry and the electronic structure of the adsorbate are shown to be important in interpreting Raman spectra to determine molecular speciation.

Raman scattering is a result of the electric field of incident radiation interacting with the electronic levels of a molecule. Valence electronic levels are not necessarily populated before adsorption and/or during irradiation. However, optically monitored properties of an adsorbate are correctly seen only as the result of properties of an adsorbate–substrate complex, in interaction with either single-particle or collective electronic properties of the substrate.

For example, the valence electron levels of a covalently bonded adsorbate interact with electron levels of the substrate to form bonding and antibonding states. The highest occupied 5σ level and the lowest unoccupied 2 pi level of the CO molecule are covalently bonded to *d* and *s* orbitals of Ni with a partial charge transfer from the donor CO to the acceptor Ni (1). This bonding causes admixed adsorbate/substrate energy levels. As a consequence, valence electron excitation by means of irradiation at an appropriate frequency range would include bonding–antibonding (intermolecular), charge-transfer, and intramolecular excitations. In the case of CN^- adsorption—CN^- being isoelectronic with CO—bonding will be partially covalent and partially ionic, depending upon the type of surface with which it interacts. Here, the surface complex would entail a transfer of charge from the donor (metal) to the acceptor (CN); subsequent irradiation of this complex can enable the monitoring of intermolecular excitations that involve the transfer of charge in the reverse direction, or photoemission that involves electron transfer from the adsorbate to the metal. For organic adsorbates such as pyridine, benzene, and so on, bonding to a surface will involve Van der Waals and covalent bonding via pi electrons of the organic adsorbate. The corresponding excitation spectra of such organic–substrate complexes will contain essentially "free" molecule and charge transfer excitations.

In general, there is a lack of information on the electronic structure and valence electron excitations (including their energies, widths, and strengths) of molecules, whether adsorbed or in solution. Specialized, but not uncommon, Raman scattering techniques allow these valence electron levels to be analyzed to a much greater extent than most other spectroscopic techniques permit, with the analysis performed in environments typical of reaction applications. Normal Raman (NR) scattering is not generally used to delineate electronic structure. With the wide variability of laser excitation frequencies now available, preresonance and resonance Raman scattering (RRS) has become more widely practiced, and it can be expected to be an increasingly important tool for probing the electronic energy levels of molecules in excited states. Brotman, Burstein, and Jiang (2) and Morris and Woodruff (3) recently discussed the application of RRS and time resolved Raman spectroscopy techniques to the analysis of the valence electronic levels of adsorbed molecules. Before an in-depth understanding of such energy levels can be obtained and subsequently used to identify the type and/or symmetry of adsorbed species, a substantial amount of background data must be accumulated and interpreted. Such data are necessary because vibrational spectroscopy relies on a comparative understanding of vibrational modes, symmetry, and energies. Historically, advances in infrared and NR spectroscopy have developed in such a comparative manner. Hence, the following information describes some NR and RRS experiments and molecules on which

NR and resonance Raman spectral measurements have been performed. Molecules in solution will be emphasized.

NORMAL RAMAN INVESTIGATIONS OF SOLUTIONS

The identification of molecular entities in solution does not necessarily entail difficult experimental procedures. The cells and instrumentation described in Chapter 3 for containing solutions and acquiring spectra are of relatively simple design. Interpretation of the spectra can be difficult, however; so spectral interpretation is emphasized here.

Small Molecules

The number of vibrational modes that are possible for a molecule is $3N - 6$, or $3N - 5$ for diatomic molecules, where N is the number of atoms in the molecule. Only a fraction of these $3N - 6$ or $3N - 5$ vibrations are Raman active. For homolytic diatomic molecules (symmetry $D_{\infty h}$), only one vibration is Raman active, its frequency depending on the mass of the atoms. For example, the vibrational frequency of $^{16}O_2$ is 1580 cm^{-1}, and for $^{14}N_2$ it is 2360 cm^{-1}. The effect of charge on such diatomic molecules can be observed by the frequencies in the series (4):

$$O_2^+ \ (1865\ cm^{-1}) > O_2\ (1580\ cm^{-1}) > O_2^- \ (1097\ cm^{-1})$$

The vibrational frequency of O_2^+ is larger than that for O_2 because O_2^+ is formed by the loss of one electron from an antibonding orbital of O_2; similarly, the frequency of O_2^- is smaller than that of O_2 because an additional electron occupies an antibonding orbital of O_2 to form O_2^-. The effect of charge also is observed for frequencies of the $C_{\infty v}$ species: $NO^+ > NO > NO^- > NO^{2-}$; such trends in vibrational frequencies are applicable in determining the nature of metal ligand species.

The vibrational frequencies of gaseous halogen species have been studied thoroughly. Upon their incorporation into organic solvents, organic molecular halogen compounds can be formed. Independent of such formation, the frequencies of the homolytic molecules can decrease as a result of charge-transfer reactions with the solvent. For example, gaseous I_2 has its vibrational mode at 213 cm^{-1}, which is decreased to 201 cm^{-1} in benzene (5,6). In aqueous solutions, the existence of diatomic units such as monomeric $[TiO \cdot aq]^{2+}$ recently has been demonstrated (7).

For triatomic molecules a number of configurations and atomic substitutions can occur. Linear X_3 and YXY molecules ($D_{\infty h}$ symmetry) have only one Raman active vibrational band. These and other molecules containing a center of symmetry have mutually exclusive infrared and Raman vibrational activity; that is, vibrations active in the Raman spectra will not be active in corresponding infrared spectra. Mutual exclusion is of general value in interpreting vibrational spectra with respect to molecular symmetry. Other than simple X_3 or YXY molecules, other molecules, such as $Ag(CN)_2$, also can contain the $D_{\infty h}$ symmetry. Here, two intense Raman bands at 2142 cm^{-1} (C—N stretching) and at 242 cm^{-1} (Ag—CN deformation) have been observed for $Ag(CN)_2$ in aqueous S_2O_3 solutions (8), whereas other Raman bands are generally weak in intensity (9). These results show that the interaction of AgCN with thiosulfate ions when present in a 1 : 1 molar ratio can be represented by:

$$2\ AgCN_{(solid)} + 2\ S_2O_3^{2-} \rightarrow Ag\ (CN)_{2\ (aq)}^{-} + Ag\ (S_2O_3)_{(aq)}$$

The symmetry of the thiosulfate ion is C_{3v} (see below), depicted as:

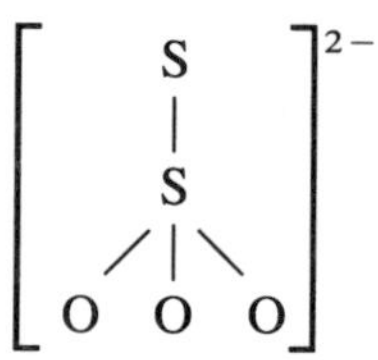

It has six fundamental modes, $3A + 3E$, all of which are infrared and Raman active, including, for example: S—O symmetric stretching at 1000 cm^{-1} ($\nu_1(A_1)$), S—O symmetric bending at 670 cm^{-1} ($\nu_2(A_1)$), S—S symmetric stretching at 450 cm^{-1} ($\nu_3(A_1)$), S—O antisymmetric stretching at 1125 cm^{-1} ($\nu_4(E)$), S—O antisymmetric bending at 535 cm^{-1} ($\nu_5(E)$), and S—S—O antisymmetric bending at 338 cm^{-1} ($\nu_6(E)$) (8). Figure 6.1, for example, shows the Raman spectrum of a 0.3 M aqueous solution of NaS_2O_3 (8). The type of coordination to the thiosulfate will change these vibrational frequencies, with shifts of the symmetric and antisymmetric S—O modes to higher values upon S-coordination and to lower values upon O-coordination. Simultaneously, the frequency of the symmetric S—S stretching mode will shift to lower values upon S-coordination (10). Raman data for several aqueous thiosulfate complexes (11) show that the antisymmetric S—O stretching mode appears at values greater than 1175 cm^{-1} for S-bridging systems, near 1130 to 1175 cm^{-1} for metal—S coordination, and at values of less than 1130 cm^{-1} for metal—O coordination.

In linear XYZ molecules ($C_{\infty v}$ symmetry) and bent X_3 and YXY (symmetry C_{2v}), the decreased molecular symmetry as compared to X_3 or YXY induces

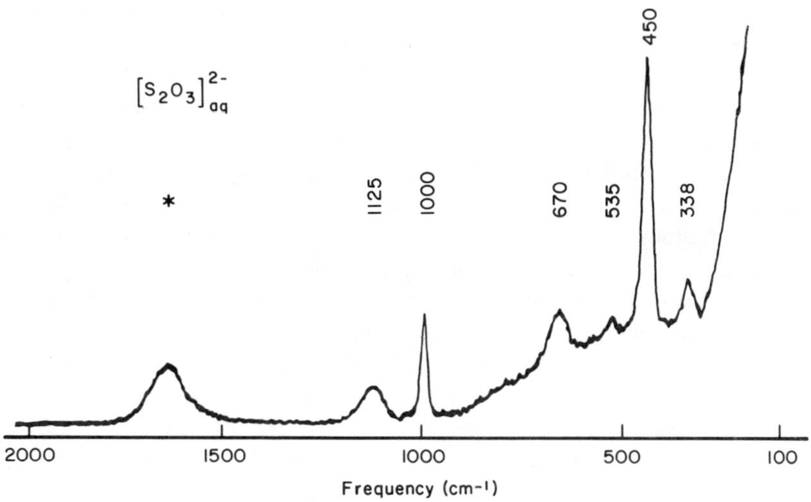

Figure 6.1. Raman spectrum of a 0.3 *m* aqueous solution of $Na_2S_2O_3 \cdot 5H_2O$ (* is an H_2O band). (8). (Reprinted by permission of John Wiley & Sons, Ltd.)

Raman allowed activity in all three of the normal vibrational modes. The angle within a bent molecule of the type YXY can be calculated by use of isotropic substitution at the X site and acquisition of the ν_3 vibrational frequency. The equation used to calculate this angle is (12):

$$[\nu(X1)3/\nu(X)_3]^2 = [M(X1)/M(X)]\,[M(X1) + 2M(Y)\sin^2\theta)]/[M(X) + 2M(Y)\sin^2\theta]$$

where M(X1) is the mass of the isotope of X and θ is the angle of bending.

The four-atom molecule of general formula XY_3 can have a pyramidal configuration (C_{3v} symmetry) or a planar configuration (D_{3h} symmetry). An important molecule in catalysis investigations that has the former symmetry is NH_3; its vibrational modes are all Raman active and in the gaseous state can contain doublets as a result of inversion doubling. The NH_3 vibrational frequencies are generally lower in metal-NH_3 and solution phase complexes than in the corresponding free NH_3 molecule (13). Upon coordination, the N—H bond weakens, causing a lower N—H stretching frequency. Stronger metal-NH_3 bond strengths further lower N—H stretching frequencies, thereby enabling an estimation of metal—NH_3 bond strengths by monitoring of such N—H vibrational frequencies. In addition to coordination, counter-ions in solution can associate by H-bonding to the N—H. In contrast, the positions of NH_3 deformation and rocking modes increase as a consequence of the formation of coordination complexes.

A planar XY_3 aqueous ion is NO_3^-. In coordination with metal cations, it is an important catalyst preparation agent for the synthesis of various oxide-supported metal and metal oxide catalysts. Investigators using Raman spectroscopy and X-ray diffraction have uniquely determined the symmetry and dimeric form of $Rh(NO_3)_3$ in aqueous solutions, as represented in Figure 6.2 (14). In general, Raman spectroscopy has been used successfully to determine ionic symmetry and the degree of anion–solvent and cation–anion interactions in NO_3^- aqueous solutions (15,16). It is anticipated that additional Raman spectroscopic investigations of other metal nitrate aqueous species would provide insight into the precursor states during catalyst preparation. For example, changes in the vibrational frequencies and multiplicities of nitrate bands can determine or help to determine the exact form of the precursor species. For solvated nitrate ions, the symmetric N—O stretching frequency is at 1052 cm^{-1} while the antisymmetric mode is at 727 cm^{-1}. Effects of the interaction of solvents and/or supports with the nitrate ion can be easily observed by Raman spectroscopy (16–18). Such polarization effects generally cause asymmetry in the low frequency side of the symmetric N—O stretching mode (ν_1) near 1050 cm^{-1}, activate the normally Raman inactive ν_2 mode near 830 cm^{-1}, and increase the separation of the ν_3 Raman active modes. Positions and assignments of the Raman vibrational frequencies of aqueous rhodium nitrate are shown in Table 6.1.

Changes in the Raman vibrational spectrum of the nitrate ion as a function of temperature of the solution have been investigated by Irish and Jarv (19) and Bulmer, Irish and Odberg (20). As the temperature of the solvent increases, its dielectric constant decreases, thereby causing weaker attractive forces between cations and anions (19,20). The positions of the nitrate vibrational bands, and the appearance of new bands due to hexaaqua–cation compounds, detail the displacement of the anion from the first coordination sphere of the cation. For aqueous $ZnNO_3$ solutions (19), increasing the temperature from 25 to 300°C

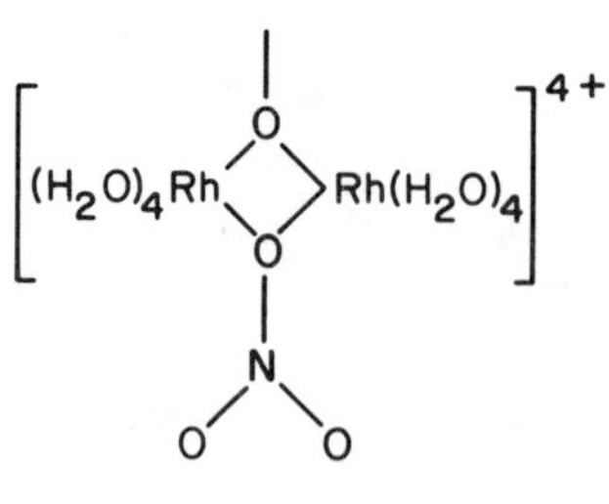

Figure 6.2. Diagram of dimeric Rh $(NO_3)_3$ in aqueous solution (14). (Reprinted with permission from *The Journal of Physical Chemistry*, Copyright 1985 American Chemical Society.)

Table 6.1 Raman frequencies (cm^{-1}) of rhodium aqueous solutions.

Rh OCTAHEDRAL CONFIG.		FREE NITRATE		BOUND NITRATE
370	ν_5	727	ν_4	781
527	ν_1	1052	ν_1	
		1348, 1423	ν_3	1302, 1525

produced an enhanced separation in the ν_3 (NO_3) stretching modes, as shown in Figure 6.3. This change was interpreted in terms of two different sites for the NO_3, one of which was in contact with the cation and other associated with H_2O molecules. In addition, Raman spectral changes as a function of temperature indicated that the octahedral configuration around Zn^{2+} decreased to tetrahedral as the temperature of the solution was increased. The nitrate ion also has been investigated in other solvents, including methanol, acetonitrile, and *p*-dioxane, by means of Raman spectroscopy (20–22). Such spectroscopic data should be useful in catalyst preparation and in delineating precursors for supported surface species.

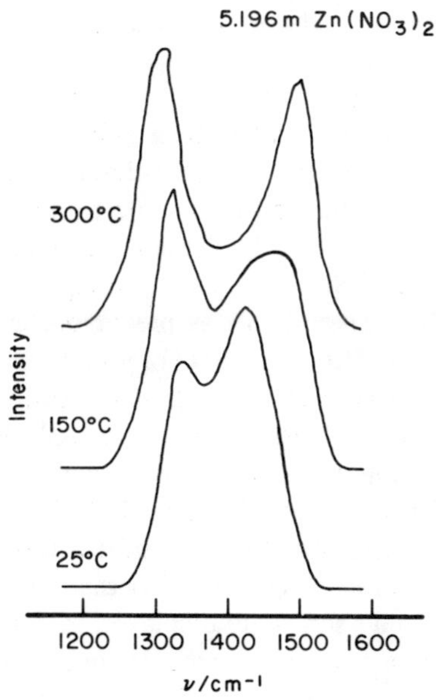

Figure 6.3. The ν_3 spectral region of nitrate ion at temperatures of 25, 150, and 300°C (19).

Besides investigating spectral changes of the solute to define cation–anion interactions and solute symmetry and form, researchers can use changes in the vibrational properties of the solvent to detail important features of species in solution. For example, Kanno and Hiraishi (23–25) showed that changes in the O—H stretching and bending vibrational regions as a function of the type of anion in aqueous Mg salts were a result of varying the strength of hydrogen bonds between water molecules and the anion. Spectral characteristics of normal and deuterated methanol and methanol–water solutions also have been used to define symmetry and solvation properties of metal salts (26–28).

A pyramidal species of interest in solution chemistry is the hydronium (H_3O^+) ion. It contains characteristic vibrational modes that can be monitored and have been used to elucidate the solution phase of $H_2PtCl_6 \cdot 2H_2O$ (29). Increasing the types of atoms in a four-atom molecule will decrease its symmetry to C_{2v}, C_1, or C_s, each of which is a subgroup of C_{3v} and D_{3h}.

All of the normal vibrational modes of the five-atom XY_4 molecule (T_d symmetry) are Raman active. In general, compounds containing XH_4 species will have decreasing XH stretching frequencies as the mass of atom X is increased. The tetrahedral symmetry is found not only in isolated or independent molecules but also in parts of larger molecules. For example, $[(CH_3)_2CHNH_3]_2[CuCl_4]$ contains tetrahedral $CuCl_4$. At relatively high pressure, the tetrahedral symmetry can be distorted, creating square-planar symmetry (D_{4h}). In solution, the interaction of tetrahedral units with the solvent can cause symmetry lowering, with concomitant creation of additional vibrational bands; this has been observed as a result of NH—Cl hydrogen bonding (30). The normal coordinate analysis of a large series of tetrahedral species has been published by Basile et al. (31).

Description of the normal coordinate vibrational bands of tetrahedral species as ν_1, ν_2, ν_3, and ν_4 allows the following rule to be formulated: $\nu_3 > \nu_1$ and $\nu_4 > \nu_2$. The intensity of ν_2 should be greater than that of ν_4, whereas solution species such as MoO_4^{2-} and ReO_4^- contain overlapping ν_4 and ν_2 frequencies. Tetrahedral species play a pivotal role as precursors in heterogeneous catalyst preparation (see Chapter 4). The assignment of vibrational bands to such a species has relied on fundamental studies of solution spectra, such as those described by Vilchez and Griffith (32), Busey and Keller (33), Woodward and Roberts (34), and others. For example, the vibrational activity of aqueous solutions of $[MnO_4]^{2-}$, $[RuO_4]^{2-}$, $[VO_4]^{3-}$, $[CrO_4]^{3-}$, and $[FeO_4]^{2-}$ and of the salts of tetrahedral species containing Zr, Mo, W, Os, and Re were elucidated by Vilchez and Griffith (32). The ratio of $\nu_1(A_1)/\nu_3(F_2)$ that they obtained for aqueous solutions of the isoelectronic series are displayed in Figure 6.4. An increase of the charge on the central atom causes a regular decrease in this ratio.

The symmetric stretching vibration of SO_4^{2-}, WO_4^{2-}, and MoO_4^{2-} anions in solution and the effect of different cations (Li^+, Na^+, K^+, Rb^+, and Cs^+) on the frequency of this vibration were investigated by Dean and Wilkinson (35);

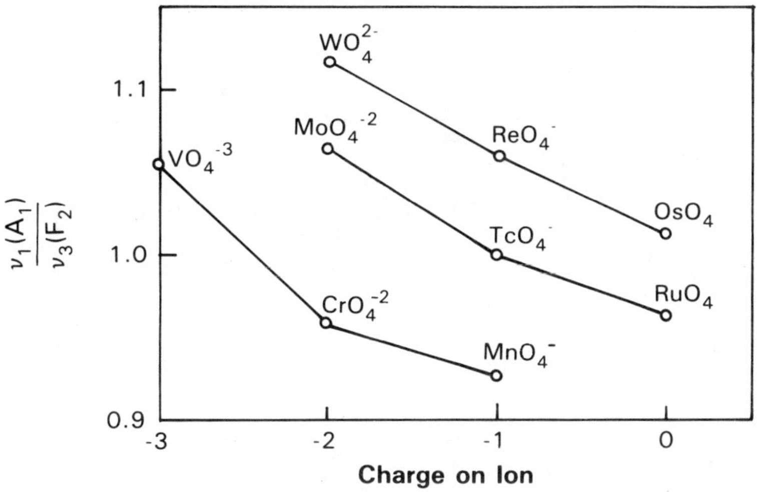

Figure 6.4. The ratio of $\nu_1(A_1)/\nu_3(F_2)$ versus the charge on ions for several oxyanions. The point for TcO_4- (aq) assumes that $\nu_1 = \nu_3$. The lines join isoelectronic species (33).

Figure 6.5 shows some of these data. The positions of the ν_1 vibrations, when plotted as a function of the cation radius for any particular concentration, were found to decrease linearly with increasing cation size over the range of Li^+ (0.06 nm) to Cs^+ (0.167 nm). It was determined that a cation of radius 0.115 nm and containing a polarizability of 0.4 Å^3 would not perturb the cations. It was suggested that when water molecules are in contact with anions in solution, the water behaves as though it has an effective radius of 0.115 nm and a polarizability of 0.4 Å^3.

Himeno and Hasegawa (36) used $Na_2MoO_4 \cdot 2H_2O$ in HNO_3, HCl, $HClO_4$, and H_2SO_4 to ascertain the presence of dimeric Mo(VI) species in solution. The principal species in these acid solutions was represented by;

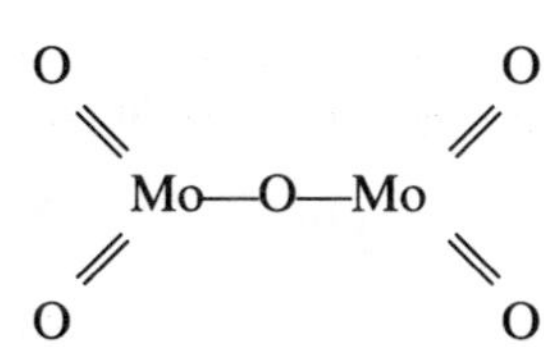

and had the vibrational frequencies that are listed in Table 6.2. The terminal vibrational frequencies, denoted by $\nu(MoO_2)$ in Table 6.2, are close in value to those determined for the Mo—O vibrational frequencies of supported molybdena

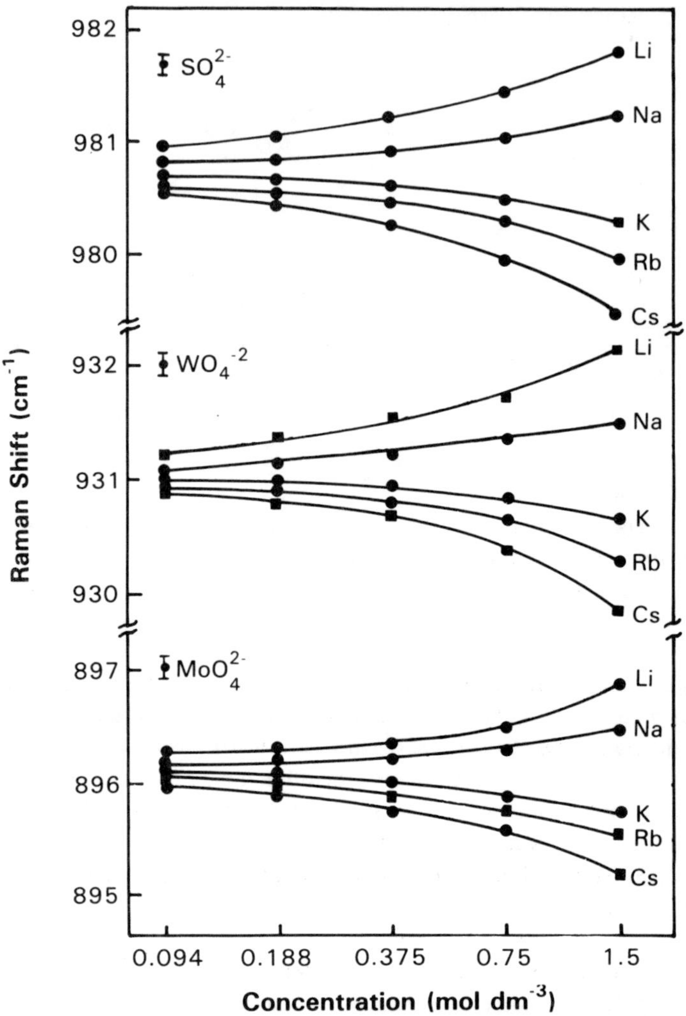

Figure 6.5. The ν_1 Raman line frequencies of anions, in solution with the alkali metal cations, as a function of the solution concentration (35). (Reprinted by permission of John Wiley & Sons, Ltd.)

species at low surface coverages (see Chapter 4). Extending these types of solution phase investigations, Himeno and coworkers (37,38) examined the Raman spectra of Mo(V), Mo(IV), and Mo(III) species in acidic solutions. The vibrational frequencies of these species are displayed in Tables 6.3 and 6.4. The Mo(IV) species was reported to occur as a trinuclear species (Mo_3) that retained

Table 6.2 Vibrational frequenices (cm^{-1}) of Mo(VI) in various acids (36).

	$\nu(MoO_2)$	$\nu(Mo_2O)$	$\delta(MoO_2)$	ν(Mo—Cl)	ν(O . . . H—O)	
HNO_3						
2.5—6 *M*	950 918	820 (720) 450	369		216	
$HClO_4$						
3 *M*	953 (937)	825 (631) (465)	378		220	
H_2SO_4						
3 *M*	951 917 (903)	830 (598) (435)	379		228	
HCl						
3 *M*	949 918	830 455	373	291	215	
conc.	957 917		387	307	213	245

a $Mo_3O_4^{4+}$ core in solution, whereas the Mo(III) species contained both trinuclear and binuclear species. A discussion of solution phase molybdenum and tungsten Raman results (39) should be useful in delineating catalyst surface speciation as a function of catalyst preparation.

The tetrahedral chlorate ion, ClO_4^-, and sulfate ion, SO_4^{2-}, have been studied by Raman spectroscopy to ascertain their association with hydrogen and the effects of concentration upon OH vibrational modes (23,28,40). Distortion of one of the Cl—O bonds would decrease the ionic symmetry from T_d to C_{3v}, equal distortion of two of the Cl—O bonds would produce C_{2v} symmetry, unequal distortion of two of the Cl—O bonds would create C_s symmetry, and any other distortion would lead to C_1 symmetry. These changes do not significantly alter the number of Raman active bands but do alter the polarizability of the bands and increase the number of infrared active bands. In the case of sulfate, a 980 cm^{-1} band was assigned to "free" SO_4^{2-}, whereas a band at 1053 cm^{-1} was attributed to "free" HSO_4^-.

Substitution of one of the Y atoms in XY_4 by a Z atom, to form XY_3Z, also lowers the molecular symmetry to C_{3v}. The presence of two Z atoms produces a molecule of C_{2v} symmetry. Incorporation of atoms other than XYZ then lowers the molecular symmetry to C_1. All of these symmetry reductions change the vibrational activity, with the correlation of the normal modes of vibration depicted in Table 6.5.

Table 6.3 Vibrational frequenices (cm^{-1}) of Mo(V) in HCl (37).

HCl/M	ν(Mo═O)	ν(Mo—O)	δ(O═Mo—O)	ν(Mo—Cl)	δ(Mo—O—Mo)	δ(Mo—O)
1	928	742 532	372	323	253	
3–5	984	870 802 436	378	326		
8–conc.	997			334		239

Table 6.4 Vibrational frequencies (cm^{-1}) of Mo_3(IV), Mo_3(III), and Mo_2(III) in HCl solutions (38).

Mo_3(IV)				
1.0–10.3 *M*	790	550–350		
Mo_3(III)				
0.5M	790	550–350		
1.0–6.0 *M*	790	550–350		285
Mo_2(III)				
1 *M*	740	500	374	
3–8 *M*	740	500	374	283

The XY_5 molecules can have D_{3h} or C_{4v} symmetry. In the case of C_{4v}, the molecule would be square-pyramidal in shape, whereas in the case of D_{3h} the molecule would be trigonal-bipyramidal. Numerous compounds of the form of $[AucL_4]^-$, $[PtCl_4]^{-2}$, $[PdCl_4]^{2-}$, and so on, of C_{4v} symmetry, have been investigated by use of vibrational spectroscopy (41,42). Examples of trigonal bipyramidal include solution phase molecules such as $[SiF_5]^-$ and $[SiCl_5]^-$ (43), and PCl_5 (44).

The XY_6 molecules belong to the octahedral (O_h) symmetry group. The Raman active modes of these molecules are labeled as the totally symmetric vibrations $\nu_1(A_{1g})$ and double and triply degenerate vibrations, $\nu_2(E_g)$ and $\nu_5(F_{1g})$. For a particular X atom, the effect of increasing the oxidation state of the molecule would be generally to increase the ν_1 symmetric stretching frequency. Many metal chloride complexes have this symmetry—[Ni, Ru, Rh, Os, Pt, Ptd, or Re] · Cl_6 compounds (45–47). In terms of relative intensities of the various Raman bands, the trend $I(\nu_1) > I(\nu_2) > I(\nu_5)$ generally holds (48). However, for $[PtCl_6]^{2-}$ and $[PdCl_6]^{2-}$ the influence of nonbonding *d* electrons in the valence shell has been described to cause $I(\nu_2) > I(\nu_1)$ (49). Alteration of the metal coordination in metal chlorides as a function of the temperature of the solution has been observed by Raman spectroscopy. For example, tetrahedral $NiCl_4^{2-}$ has been shown to be favored over octahedral $NiCl_6^{4-}$ at temperatures above

Table 6.5 Correlation of the vibrational modes for the symmetry species T_d, C_{3v}, C_{2b}, and C_1.

GROUP	ν_1	ν_2	ν_3	ν_4
T_d	A_1	E	F_2	F_2
C_{3v}	A_1	E	$A_1 + E$	$A_1 + E$
C_{2v}	A_1	$A_1 + A_2$	$A_1 + B_1 + B_2$	$A_1 + B_1 + B_2$
C_1	A	$2A$	$3A$	$3A$

180°C (50). Raman spectroscopy also can provide a measurement of enthalpy changes associated with octahedral to tetrahedral transformations, in, for example (51):

$$Co(H_2O)_6^{2+} \rightleftarrows Co(H_2O)_4^{2+} + 2H_2O$$

Discrimination between the various symmetries of X_2Y_4 molecules by Raman spectroscopy should be somewhat easy. For example, the symmetry groups to which this type of molecule can belong include D_{2h}, D_{2d}, and D_2. The D_{2h} symmetry would involve a planar molecule with a center of symmetry. For D_{2d} there should be 9 Raman active vibrational modes, whereas for D_2 there should be 12 Raman active modes. Bridged X_2Y_6 molecules also have D_{2h} symmetry (52). The symmetry of ethane-type X_2Y_6 molecules is D_{3d}. This nonbridged species can be distinguished from bridged species because it will contain a lower number of Raman active bands than will the bridged species.

Molecules of the type X_2Y_7 belong to the groups C_s, C_{2v}, or C_1 depending on the relative orientation of the XY_3 groups. This type of molecule has been intensively studied by Griffith and Lesniak (53) and Brown and Ross (54). Vibrational spectra of a $[V_2O_7]^{4-}$ complex have been observed in dual function, fluid catalytic cracking (DFFCC) catalysts after steam treatments (55); its presence has been used to delineate FCC catalyst deactivation mechanisms. Of course, the type of vanadia species that would be present in solution depends upon the pH of the solution, as suggested by Griffith and coworkers (53,56), and given by:

$$(n/2)(V_2O_7)^{4-} + nH^+ \rightleftarrows (VO_3)_n^{n-} + (n/2)H_2O$$

The surface speciation of supported Re_2O_7 metathesis catalysts have been studied by Raman spectroscopy (see Chapter 4). It is expected that in the solid state rhenate would be polymeric in nature, whereas it would be monomeric in the liquid state (57).

Complex Molecules

Chassaing, Basile, and Lordthioir (58) examined Ti(III) solutions in alkali metal chlorides and mixtures and showed that the solvent cations determined the type of anions formed. The Raman spectra of tetrhedral $TiCl_4^-$ (59) and octahedral $TiCl_6^{3-}$ (58–61) in solution were examined as a function of the concentration of LiCl. A new species was formed at high Cl^- concentrations that was described as $Ti_2Cl_9^{3-}$. The interpretation of these spectra may be important in the delinea-

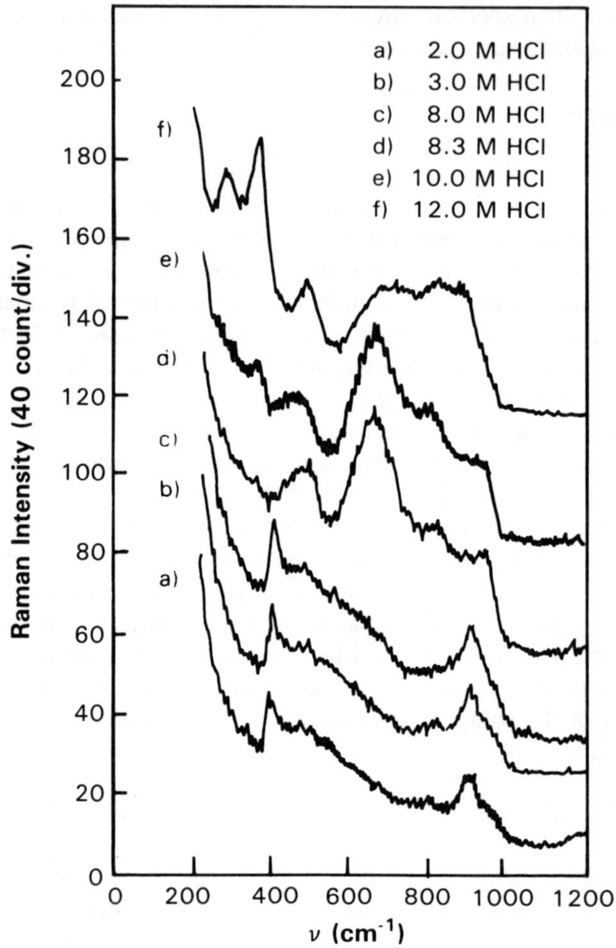

Figure 6.6. Raman spectra of $TiCl_4$ solutions as a function of HCl concentration (62). (Reprinted with permission from *Langmuir*. Copyright 1987 American Chemical Society.)

tion of surface speciation observed by Reichmann and Bell (62) for $[Ti_8O_{12}(H_2O)_{24}]Cl_8 \cdot HCl \cdot 7H_2O$. Here, Reichmann and Bell (62) assigned the complex Raman spectra by comparison with the Raman spectra of $TiCl_4$ in aqueous HCl solutions. The Raman spectra of these solutions are depicted in Figure 6.6. Substitution of D_2O for H_2O did not change the vibrational characteristics in these spectra. Hence, most of the bands observed in Figure 6.6

were assigned to Ti—O—Ti bridging vibrational modes; the symmetry of the solution species was not described.

Carbonyls

Mononuclear and polynuclear carbonyls have been used extensively in the preparation of catalysts and, correspondingly, have been extensively studied by vibrational techniques to ascertain their structure. Jones and coworkers (63,64) have performed extensive normal coordinate and solution phase spectral analyses of these carbonyls. The M—C bond strength analysis for the carbonyls containing Mo, Cr, W, and Re, estimated by Raman analysis, has not been completely satisfactory (65,66). The Raman spectra of $Os_3(CO)_{12}$ and $Ru_3(CO)_{12}$ were investigated by Quicksall and Spiro (67). Six carbonyl bands are expected for these compounds; for $Os_3(CO)_{12}$, six bands were observed, the positions of which were 2130 (A_1'), 2028 (E''), 2019 (E'), 2006 $(A_1)'$, 2000 (E'), and 1989 (E') cm^{-1}. Spectral resolution of these bands could not have been accomplished by use of infrared spectroscopic techniques.

The metal–metal stretching vibrations of a number of mononuclear and polynuclear metal complexes have been assigned. For binuclear carbonyls, positions of metal–metal bands were 160 cm^{-1} for Mn–Mn, 148 cm^{-1} for Tc–Tc, 122 cm^{-1} for Re–Re (68), and 178 cm^{-1} for Pt–Pt (69). However, the observation of relatively intense bands in the 100 to 250 cm^{-1} region is not always indicative of metal–metal stretching vibrations. For example, the proposed assignment of 225 and 219 cm^{-1} bands in the Raman spectra of $Fe_2(CO)_9$ and $Fe_3(CO)_{12}$ has been determined to be a result of decomposition of these complexes upon laser irradiation (70).

The compounds $Re_2Cl_8^{2-}$, $Mo_2Cl_8^{4-}$, $Mo_2(OAc)_4$, and so on, do not contain carbonyl functionality but do contain metal–metal bonding. Dependence of the metal–metal vibrational frequencies on the metal–metal bond order has been identified in the case of Mo-, Re-, Mn-, Bi-, and Pb-binuclear compounds (71,72). The ability of Raman techniques to measure these low frequency modes is a direct result of usually minimal interferences in this low frequency region and of the relatively high spectral resolution obtainable.

Hydrocarbonyls contain metal—H and CO bands located close to each other in the 2200 to 1800 cm^{-1} region. However, deuteration of the hydrocarbonyls and the relatively high spectral resolution available with Raman spectroscopy allow bands associated with these species to be distinguished. A complete vibrational analysis for $HMn(CO)_5$ was performed by Edgell et al. (73). Terminal metal—H vibrational bands were located nearly 1000 cm^{-1} above bridging metal—H bands. For example, for $FeHCo_3(CO)_{12}$ the bridging metal—H band was located at 1114 cm^{-1} (74), whereas in $Re_2H_2(CO)_8$ the Re—H bridging species contained bridging metal—H bands at 1382 and 1272 cm^{-1} (75).

Acetic Acid

Besides identifying the form of molecules in solution and the effects of solvents on solutes, Raman spectroscopy can be used for discrimination of monomeric and dimeric species in solution. A good example of this type of research is found in a recent publication by Ng and Shurvell (76). Here, the monomer–cyclic dimer equilibrium in acetic acid–water solutions was examined by a detailed spectroscopic analysis of band contours in the carbonyl stretching region. A considerable amount of H-bonding is present in such a system; deconvolution of bands that are broadened by H-bonding effects can be accomplished by factor analysis and (77,78) in conjunction with quantitation methods (internal stan-

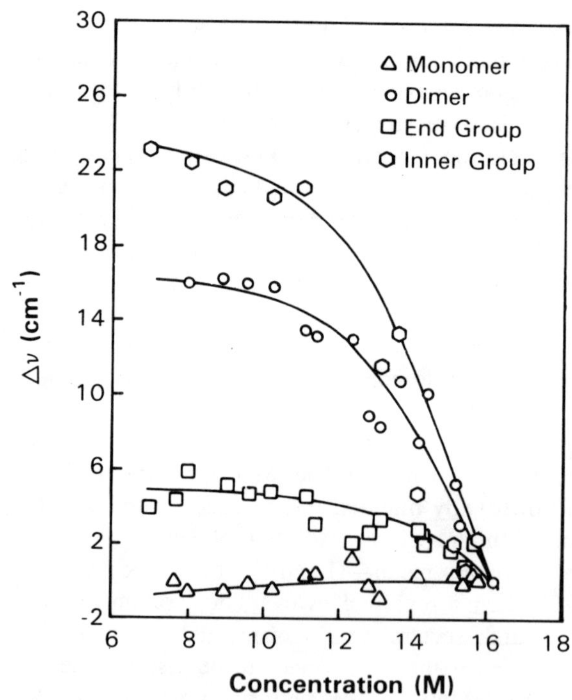

Figure 6.7. Frequency shifts of C = O bands (relative to the positions in 16.2 *M* CH_3COOH) (76). (Reprinted with permission from *The Journal of Physical Chemistry*, Copyright 1987 American Chemical Society.)

dards). Figure 6.7 depicts the change of C═O bands associated with particular species as a function of increasing CH_3COOH concentration. The vibrational frequencies of the end CO group and the monomer are not affected as much as those of the dimer and the inner CO group. The bending mode of H_2O also was shown to decrease with increasing acetic acid concentration.

Heteropoly Anions

Heteropoly compounds recently have received attention for their unique catalytic properties (79–82). These compounds are ionic solids with complex, high molecular weight anions consisting of a cagelike structure, often referred to as the Keggin unit (83). The central metal is typically Si, As, or P, whereas the peripheral atoms are generally W, Mo, V, and Ti. There have been numerous Raman studies identifying the pH dependence of heteropoly anion decomposition and the presence of various heteropoly species in solution (84,85). Because these compounds have been used in redox catalysis, their reduced and oxidized states are of interest. For molybdo- and tungsto-heteropoly anions, the structures of the oxidized and reduced state have been assumed to be identical, even though visible absorption characteristics of these states are not identical (86). Raman results that confirm this assumption are shown in Figure 6.8 and band frequencies and intensities in Table 6.6. The Raman spectra are identical, independent of

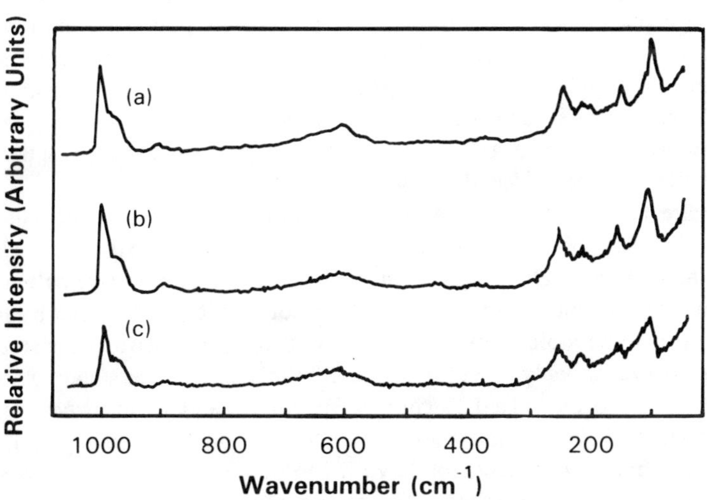

Figure 6.8. Raman spectra of 0.1 *M* aqueous solutions of molybdovanadophosphates using 647.1 nm excitation: (a) $PVMo_{11}O_{40}^{4-}$; (b) $PV_2MO_{10}O_{40}^{5-}$; (c) $PV_3MO_9O_{40}^{6-}$ (86).

Table 6.6 Raman band positions (cm^{-1}) and relative intensities for the molybdovanadophosphates.[a] (86)

$PVMo_{11}O_{40}^{4-}$	$PV_2Mo_{10}O_{40}^{5-}$	$PV_3Mo_9O_{40}^{6-}$
105 (6)	105 (6)	108 (7)
155 (2)	154 (2)	156 (3)
214 (2)	213 (2)	213 (2)
250 (5)	252 (5)	255 (5)
378 (1/2)	382 (1/2)	. . .
451 (1/2)	453 (1/2)	450 (1)
605 (2)	608 (2)	607 (3)
649 (1)	657 (1)	. . .
897 (1)	896 (1)	892 (1)
972 (4)	972 (5)	973 (5)
995 (10)	992 (10)	992 (10)

[a]Numbers in parentheses denote intensities relative to strongest peak.

increases in the reduced nature of the sample. These types of complex structures can be identified by acquisition of the fundamental vibrational spectra, in which the relative intensities of bands and band position are closely compared to results from complexes identified by X-ray diffraction. These Raman results can be compared to the data of Ng and Gulari (87) who determined that the aggregation scheme of molybdate solutions as a function of pH proceeded in the form of simple → hepta → octa → [$Mo_{36}O_{112}$] → protonated polymeric molybdenum species, whereas the aggregation scheme for tungstate solutions proceeded in the form of simple → Y-polytungstate → paratungstate-A → paratungstate-B → ψ-metatungstate. Identification of heteropoly compounds for supported Mo and correlation of vibrational frequencies of supported Mo and W with the above type of solution species was discussed in Chapter 4.

There are numerous other examples of simple and complex molecules that have been or could be studied by normal Raman spectroscopic methods. Cation–solvent interactions, cation–ligand interactions, cation–anion interactions, and the symmetry and state of aggregation of the solute can be identified by Raman analysis. It is impossible to list or to discuss all Raman spectroscopic studies of solutions. However, only recently has the importance of understanding solution phase chemistry been related to catalyst speciation and performance. Besides these NR analyses, recent advances in excited state electronic theory and pulsed resonance Raman experimentation have led to numerous studies that should be of value to catalysis researchers. The following section summarizes some of these results.

RESONANCE RAMAN ANALYSIS OF SOLUTIONS

As described in Chapter 3, on instrumentation, recent advances in resonant Raman techniques include the use of pulsed excitation laser sources and improved detectors having time resolution as small as 10^{-12} second. Standard notation for time-resolved resonance Raman scattering is TR^3; for time-resolved vibrational spectroscopy it is TRVS. Since the first reports of TR^3 about a decade ago (88–90) the application of TR^3 has been limited to systems on which complementary NR spectroscopic data can be obtained so that spectral interpretation can be accomplished. Although time-resolved emission and absorption are extremely sensitive and applicable to fast solution kinetics, the associated electronic spectra often contain broad and ill-defined peaks. In contrast, resonance Raman spectra can be acquired within the same time domain as time-resolved emission and absorption, but offer specific molecular-sensitive information at the relatively high spectral resolution that is inherent to Raman scattering. The use of RRS also can provide information on symmetries, energies, and chemical kinetics at low molecular concentrations. Hopefully, future TR^3 studies will focus on the identification of intermediates in chemical reactions under realistic catalytic conditions.

Wilbrandt et al. (90) were among the first to develop and use TR^3, publishing a spectrum of the relatively stable radical diphenyl-pikryl-hydrazyl at concentrations as low as 2×10^{-4} mole/liter in benzene and ethylbenzene solutions. The time resolution of their experimentation subsequent to excitation was 600 ns. Only two bands, at 1340 and 1217 cm^{-1}, were observed; that at 1340 cm^{-1} was assigned and to the N—O symmetric stretching mode of a NO_2 species, whereas the 1217 cm^{-1} band could not be firmly assigned. At about the same time, Harada et al. (91) studied the RRS of N-methylacetamide, N-ethylacetamide, and N-methylpropionamide using laser excitation in the range of 257.2 to 647.1 nm. The presence of intensity enhancements in particular vibrational modes, above that expected for a ν^4 dependence in NR spectroscopy, was particularly evident when the 257.3 nm ultraviolet excitation was used. This enhancement was attributed to resonant enhancement associated with the electronic absorption mode, $pi_1 \rightarrow pi^*$; such resonant excitation was reported to cause contraction of the C—N bond distance.

Bipyridine Complexes

Dallinger and Woodruff (92) presented the first TR^3 spectrum of an optical-ly generated excited state in solutions, in their study of tris(2,2′-bipyridine)-ruthenium(II) in its lowest excited state, $^*Ru(bpy)_3^{2+}$. Prior to this publication, excitation of molecules to higher electronic levels had been accomplished pri-

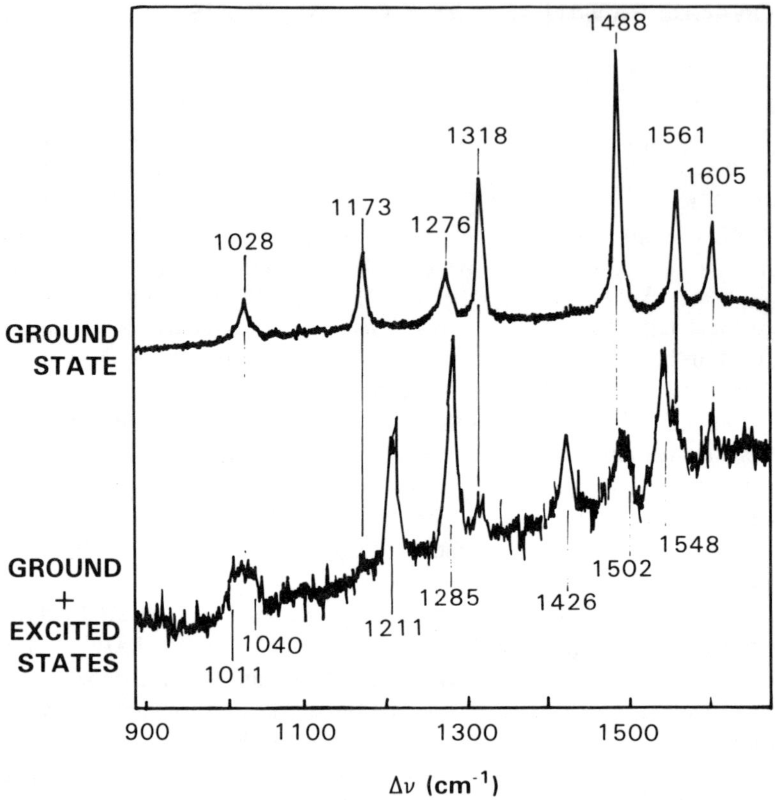

Figure 6.9. Upper trace, ground state spectrum of $Ru(bpy)_3{}^{2+}$ taken with 457.9 nm excitation. Approximately 2.5×10^{-4} *M* solution in H_2O, laser power 100 mW at the sample, spectral slit width ~ 5 cm^{-1}. Lower trace, ground plus excited state spectrum of $Ru(bpy)_3{}^{2+}$ taken with 354.4 nm pulsed excitation. Approximately 2.5×10^{-4} *M* solution in H_2O, laser operating at 15 Hz and 75 mW average power (5 mJ/pulse) (92). (Reprinted with permission from *The Journal of Physical Chemistry*. Copyright 1979 American Chemical Society.)

marily with the use of ultraviolet flash lamps. The molecule $Ru(bpy)_3^{2+}$ has been extensively studied for its potential as a photochemically active molecule (93,94), and the first excited electronic level has been found to have a measured lifetime of approximately 655 ns. Spectra of the ground state and the first excited state of this molecule are shown in Figure 6.9. The excited state spectrum contains bands in excess of those observed in the ground state, NR spectrum. Intensities of the excited state vibrational modes are nearly equal to those observed for the ground state, as a result of optical pumping of electrons from the ground state to the excited state. A direct correlation of bands in the excited and NR spectra was not entirely attempted; however, frequencies of the excited modes were, on the average, about 60 cm^{-1} lower than the NR frequencies. In addition, it was surmised that, within the time scale of the vibrational modes, the electron density within the excited state of the molecule was predominantly localized on one of the bipyridine (bpy) ligands instead of delocalized over all three ligands.

The extent of electron delocalization in the bpy ligands was addressed further in subsequently obtained TR^3 results (95). Whether the first excited state should be formulated with complete delocalization, as in $[Ru^{3+}(bpy^{-1/3})_3]^{2+*}$, or with localization on one ligand, as in $[Ru^{3+}(bpy^-)(bpy^0)_2]^{2+*}$, is of considerable importance because this excited metal-to-ligand charge transfer state is the source of the photochemical activity of this molecule. The deuterated analogue of $[Ru(bpy)_3]^{2+}$ was investigated to enable a more exact correlation of NR vibrational frequencies and assignments with those from the excited state (96). This study resulted in essentially the same conclusions as the assignments in reference 91.

Smothers and Wrighton (97) reexamined TR^3 spectra of the same $Ru(bpy)_3$ ligand, in addition to obtaining results for a mono(bpy) complex containing Re. A comparison of Raman bands for the ground and excited states for these complexes was obtained, and it showed excellent agreement between frequencies for the Re and Ru complexes, indicating that the localized model of the charge transfer state in $[Ru(bpy)_3]^{2+}$ was correct.

Bipyridine complexes of the general formula $[Ru^{2+}(bpy)_{3-n}(4,4'\text{-}(CH_3)_3)(bpy)_n]^{2+}$ were investigated with the use of TR^3 techniques by McClanahan, Hayes, and Kincaid (98). The electron was found to be localized on one of the ligands, the one that could most easily be reduced. The TR^3 results were also interpreted to show a mixture of excited state species instead of a single excited state; for example, in the case of $[Ru(bpy)(CH_3)_2\text{-}bpy)_2]^{2+}$, Raman bands were observed that indicated bpy^- and another ligand ion.

For Os–bpy complexes, Bradley et al. (99) studied the extent of transfer to the bpy ligands by quantifying in complexed bpy the shifts in vibrational frequencies between ground and excited states relative to shifts between uncomplexed bpy. Bond length displacements of the CC and CN vibrational modes in the excited state spectra also were calculated. These displacements may be as precise as results from NR spectra of condensed phases.

Casper et al. (100) also reported TR^3 spectra of Os–bpy complexes, having the general formula $[Os^{2+}(bpy)_n]((cis\text{-}1,2\text{-bis(diphenylphosphine)ethylene})_2)_{3-n}$. Ligand localization/delocalization of the charge, the extent of charge transfer, and bond displacement values were quantified. It was noted that considerable interaction occurred between the Os *d*-pi orbitals and the bpy pi* excited state orbitals.

Resonant Raman data on 1,1′-dimethyl-4,4′-bipyridine dication (MV^{2+}) was investigated by Forster, Girling, and Hester (101). This complex has been used as an electron relay in the photochemical splitting of H_2O into hydrogen and oxygen (102,103). Inorganic metal complexes, when in excited states, reduce MV^{2+} to its radical cation MV^+; in turn, MV^+ can reduce H_2O at neutral pH in the presence of noble metal catalysts (103–106). Aqueous solutions at concentrations as low as 10^{-3} *M* MV^{2+} were examined, and NR spectral results were obtained for solution and solid state samples that allowed extensive assignments of the vibrational modes. The radical cation MV^+ was prepared by addition of Zn to MV^{2+} solutions with the subsequent acquisition of RRS and NR spectra. By assignment of the vibrational bands, it was possible to distinguish the presence of a dimer—MV^+ in the MV^+ solution.

Mabrouk and Wrighton (107) studied electron transfer from fac-$ClRe(CO)_3(bpy)$ to MV^{2+} by use of TR^3 techniques. They showed the time evolution of MV^{2+} to MV^+ for the reduction of MV^{2+} by fac-$ClRe(CO)_3(bpy)$, and also obtained TR^3 results for $Ru(bpy)_3^{2+}$ that were identical to previous studies on these molecules (94,95).

Aromatic Hydrocarbons

The coupling of electronic and vibrational levels in the excited state Raman spectra of benzene and benzene-d_6 of approximately 0.1 *M* concentration in acetonitrile was theoretically and experimentally investigated by Ziegler and Hudson (108). Such vibronic coupling to the second excited singlet state ($^1B_{1u}$), activated in benzene by use of 212.8 nm radiation, produced Raman spectra dominated by overtones and combination bands from a dipole forbidden transition, E_{2g}, as well as by the fundamental, overtone, and combination bands from the A_{1g} symmetric stretching, ring breathing mode of benzene. In general, depolarization ratios and relative intensities were quantitatively explained in terms of vibronic theory (109,110). The totally symmetric ν_1 mode was the only fundamental that showed resonance enhancement in intensity. In toluene, para-xylene, and mesitylene, also excited by use of 212.8 nm radiation, the intensity of the symmetric stretching ring breathing mode was weak compared to its intensity in benzene (111).

A TR^3 study of 1,4-diazanaphthalene (quinoxaline) in aqueous solution (112,113) at concentrations as low as 10^{-3} *M* showed that transient absorption results had not been able to sufficiently delineate protonation reactions in this complex. As in the case of RRS of benzene, discussed above, sampling techniques included flowing the solution through a micro-volume cell at a rate that exchanged the sample between laser pulses. The NR (pulsed) spectrum and the corresponding TR^3 spectra are displayed in Figure 6.10; corresponding frequencies of the bands are listed in Table 6.7. Eight bands in the NR spectrum are superimposed on a broad continuum due to H_2O, with six additional bands appearing upon resonant

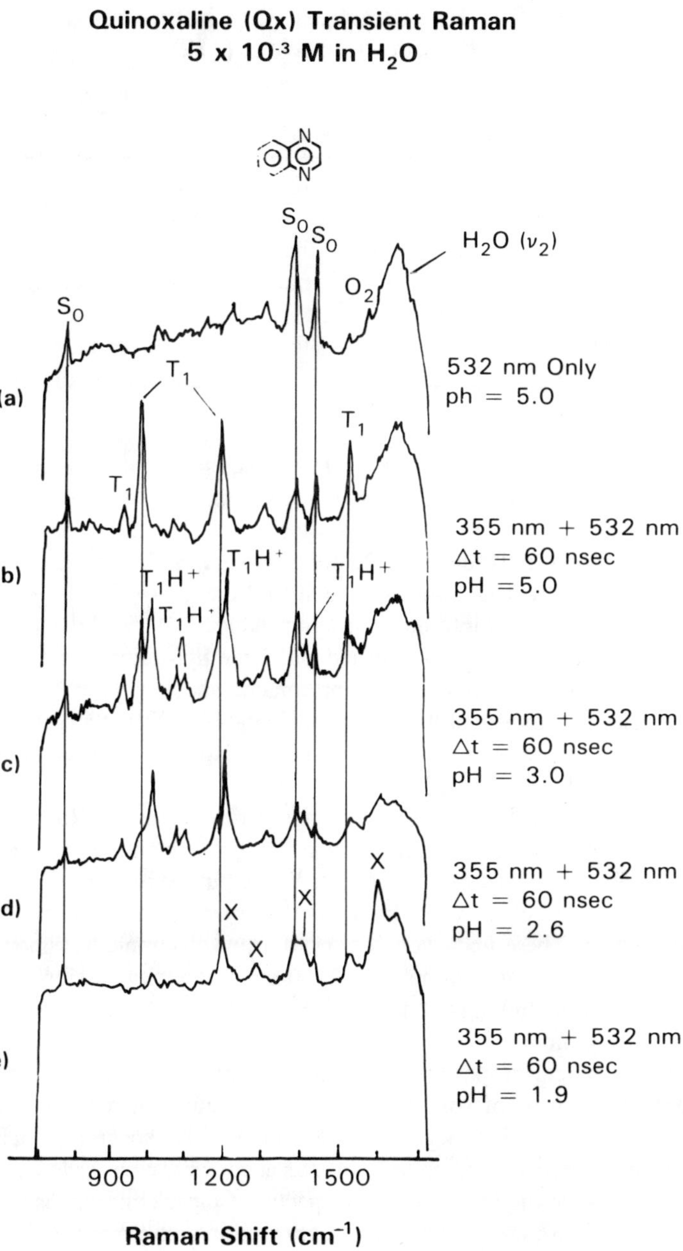

Figure 6.10. The Raman spectra of quinoxaline in H_2O as a function of pH with a change in time of $\Delta t = 60$ ns (112).

Table 6.7. Positions (cm^{-1}) of bands for the indicated species. (112)

Qx H_2^{2+} (S_0)	Qx H^+ (S_0)	Qx (S_0)	Qx (T_1) 532 nm	Qx (T_1) 416 nm
		1424	1504	1504
1535	1495	1372		1389
1498	1409	1296		1290
1470	1389	1210	1182	1182
1416	1358	1145	972	972
1395	1281	1034		
1360	1011	1017		
1275	761	768		

excitation. The triplet excited state (T_1) becomes protonated (T_1H^+) at 100 ns after the excitation pulse when the pH is greater than 5. At lower pH, the T_1H^+ products appear at smaller delay times after excitation. In all cases, the stable, and final, product was labeled X. Deuteration studies, in which D_2O was used as solvent, showed that D^+ not only catalyzed the $T_1 \rightarrow T_1H^+ \rightarrow X$ reaction, but also was incorporated into the transient product T_1H^+ and the stable product X. Analogous studies were performed on 2-methyl-quinoxaline. A tentative identification of X was given as the diprotonated species, T_1H^{2+}; such a diprotonated species was expected to have a considerably different structure from that of ground or singly protonated species (114,115). However, a structure for T_1H^{2+} was not presented.

Kobayashi and Koshihara (116) compared time-resolved Raman data with time-resolved fluorescence and absorption for diphenylamine in *n*-hexane solutions at concentrations as low as 10^{-1} *M*. Weak but reproducible bands at 1155 and 1703 cm^{-1} were observed, which were analyzed as a function of time after sample excitation. These lines were attributed to the electronic transition $T_n \leftarrow T_1$. With the use of deuterated samples, a 1150 cm^{-1} band was assigned to the C—N—C symmetric stretching mode, and the 1703 cm^{-1} band was assigned to a C—N stretching mode.

The photophysical properties of acridine in solution have been shown to be dependent on the type of solvent and the temperature of the solution. Hence, Van Zeyl, Varma, and Vroege (117) examined TR^3 spectra of acridine and anthracene in benzene and acetonitrile solvents at concentrations as low as 5×10^{-4} *M*. Assignments for the NR vibrational spectra of these compounds were readily available in the literature, and electronic absorption also was well characterized. Tripathi and Fisher (118) also studied the TR^3 behavior of anthracene and, in the process, tentatively assigned its vibrational modes using the assumption that the normal motions of the atoms do not change drastically between the ground state, S_0, and the lowest triplet state, T_1. Decay kinetics of the triplet state were evaluated.

Quantitation

Chinsky et al. (119) and Blazej and Peticolas (120) employed Kramer-Kronig (KK) transform and RRS techniques to define shifts in the excited state potential energy minimum along the normal coordinates in uracil. Only the excitation profiles for the fundamental Raman frequencies were needed to determine these shifts by use of this KK transform technique. An internal standard was not used to quantify intensities of the bands; instead, the first two overtones of each of the five fundamentals of uracil were used for standardization. This procedure was shown to decrease significantly the computation needed to quantify RRS results.

Acquisition of quantitative measurements from TR^3 data has not been thoroughly addressed. Although an internal standard method of quantitation, as previously discussed for NR spectra in Chapter 4, provides somewhat satisfactory results, the fact that resonant Raman spectroscopy relies on relatively strong absorption by the molecule under study implies that using an internal standard requires corrections for self-absorption—that is, reabsorption of the emitted Raman scattering. Such reabsorption generally is not a problem in NR scattering because in many reported cases the samples have very weak color characteristics. Theoretical investigations have been developed to handle quantification (121–123) if Frank-Condon overlap integrals can be estimated. More simplified analyses also have been described (124,125), but their precision relies on a small shift along the normal coordinates in the excited electronic states. Hence, Blazej and Peticolas (126) initiated a study on the use of excitation profiles of specific vibrational modes within the excited state spectra as a means of quantifying the measurement of TR^3 spectra. Their method was utilized in the study of adenosine 5′-monophosphate, cytidine 5′-monophosphate, and uridine 5′-monophosphate. Their parameterized method also was found useful in determining the geometry of the excited state.

The development of quantitative and semiquantitative resonant Raman methods is ongoing, not only with the goal of being able quantitatively to measure minute concentrations of transient or equilibrium species that may be involved in reactions, but also with the objective of being able to separate Raman spectra of complex mixtures. For pyrene in pure solutions, it has been shown that concentrations lower than ppb can be monitored routinely (127). Perhaps the most severe difficulties that could be encountered in identifying molecular species and measuring their relative absolute concentrations would be found in the RRS analysis of coal liquid distillates. Recently, Rumelfanger, Asher, and Perry (128) used RRS to characterize the polycyclic aromatic hydrocarbon composition in such samples. Samples were produced from distillate cuts at various temperatures of a coal-derived product. The UV absorption and resonant Raman spectra of pure compounds were compared to those obtained from the distillate cuts. The absolute concentration of pyrene as a function of distillate cut could be determined by this method, but the concentrations of other aromatic compounds were not uniquely determined because of the lack of an adequate library of resonant spectra for aromatic hydrocarbons.

Micellar Solutions

The influence of the local arrangement of solvating molecules on chemical reaction of the solute should not be overlooked. Reactions occurring in enzymes, membranes, and micelles may be particularly sensitive to such an arrangement because, within microscopic regions, hydrophobic character can exist a very short distance from hydrophilic character. For example, the interior region of a hydrocarbon micellar could be separated from an aqueous phase by an ionic double layer, creating a large potential gradient across the micelle boundary region. Molecules could be stabilized within the micelle interior and undergo destabilization at the boundary or in the aqueous phase outside the micelle. The opposite also could occur. Additionally, a hydrocarbon molecule could be dissolved within the micelle but not in the aqueous phase. Excitation of such a micelle-trapped molecule using a specifically tuned laser pulse would be capable of ionizing this trapped molecule. As a result of excitation, ejection of an electron into the boundary region or aqueous phase could occur. The resulting anionic hydrocarbon would be electrostatically attracted to the boundary region where a potential exists. Such a chemical reaction and the kinetics of reaction have been investigated by Beck and Brus (129) by the use of TR^3 techniques on aqueous micellar solutions of tetramethylbenzidine (TMB).

As TMB is hydrophobic, it dissolves within the interior of the micelle. Photolysis or oxidation causes the formation of TMB^+ and TMB^{2+}, which have intense absorption in the 400 to 500 nm spectral region (130,131), the frequency of such absorption being strongly dependent upon the micelle used to stabilize the parent TMB. Raman results for TMB^+ and TMB^{2+} are shown in Figure 6.11, with corresponding assignments listed in Table 6.8. Disproportionation of TMB:

$$TMB^+ + TMB^+ + 2H^+ \rightarrow TMB^{2+} + TMBH_2^{2+}$$

could not be positively identified in the TR^3 results, either because of possibly identical spectra of TMB^{2+} and $TMBH^{2+}$ or because $TMBH^{2+}$ was not a transient species in the reaction of $TMB^+ \rightarrow TMB^{2+}$. However, a distorted TMB^+ intermediate species was identified and shown to be solvated within the

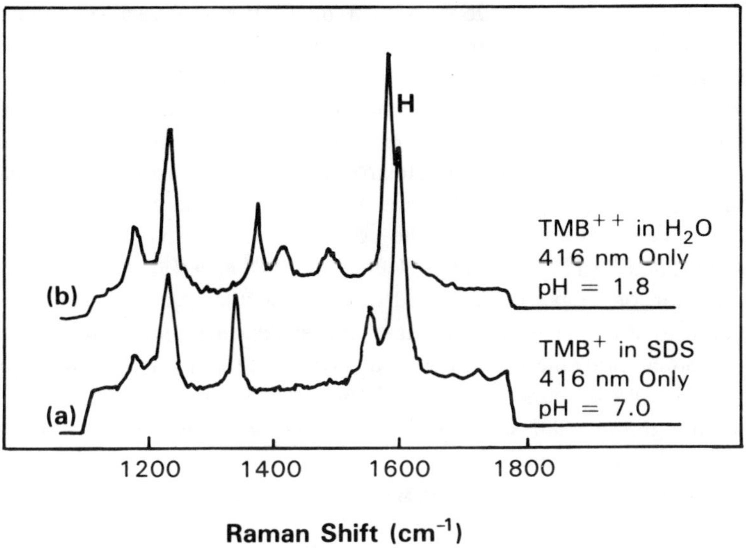

Figure 6.11. Reference Raman spectra (416 nm): (a) TMB^+ in SDS micellar solution formed by oxidation with dilute Br_2 water; (b) TMB^{3+} formed by oxidation of a $TMBH_2^{3+}$ solution (pH 1.8) with Br_2 water. Instrumental resolution = 5 cm^{-1} (129). (Reprinted with permission from the *Journal of Physical Chemistry*. Copyright 1983 American Chemical Society.)

Table 6.8 Raman frequencies (cm^{-1}) of TMB^+ and TMB^{2+}.
(Ref. 129, reprinted with permission from J. Am. Chem. Soc., *105,* 1106 (1983)

MODE TYPE	TMB^+	*I*	TMB^{2+}
Ring C—C stretch	1604	1606	1590
781 cm^{-1} overtone	1559	1561	1495
Inter-ring C—C stretch	1343	1346	1420
C—H in-plane bend	1234	1241	1372
C—H out-or-plane bend	991	991	1239
	941	941	1182
Ring breathing	781	781	927
			759
			636

double layer. Ejection of an electron from the TMB was shown to reduce MV^{2+} species, which were dissolved in the aqueous phase.

The TR^3 characterization of the molecular species created by excited state electron transfer on colloidal TiO_2 and CdS was investigated by Rossetti and coworkers (132,133). In these studies, aqueous soluble molecular precursors, bound to the colloidal surfaces by electrostatic attraction or by adsorption, were electronically excited by pulsed irradiation, thereby causing the ejection of an electron, which was captured by the colloid. The oxidized, adsorbed species was then probed by TR^3 techniques. Generally, TR^3 spectral results are sensitive to protonation or solvation of the adsorbed molecule; such sensitivity enables delineation of the local environment of the adsorbed molecule and distinction of the adsorbed molecule's symmetry and type. Raman spectra could be obtained for eosin on TiO_2 at concentrations as low as 5×10^{-5} *M* (see Figure 6.12). The parent, ground state eosin and the excited state, partially oxidized eosin were easily differentiated.

The electron and hole transfer reactions:

$$\text{hole}^2(TiO_2) + SCN \rightarrow SCN \rightarrow (SCN)_2^-$$
$$e^-(TiO_2) + MV^{2+} \rightarrow MV^+$$
$$e^-(TiO_2) + C_{14}MV^{2+} \rightarrow C_{14}MV^+$$
$$e^-(CdS) + MV^{2+} \rightarrow MV^+$$

also were investigated by Rossetti, Beck, and Brus (133,134). The TR^3 results of these reactions indicated that the products had aqueous first solvation shells within the time frame (nanoseconds) of the resonant excitation.

In an extension of their work on the reduction of MV^{2+}, Rossetti and Brus (135) used picosecond resonance Raman techniques to follow the excitation of MV^{2+} adsorbed onto CdS and ZnS colloids under conditions of S^{2-}-rich and Cd^{2+} surfaces. The differences observed in the Raman spectra from these surfaces, that is, relatively low intensity MV^+ spectra from S^{2-}-rich ZnS colloids as compared to S^{2-}-rich CdS colloids, indicated a photoreduction mechanism not recognized in previous nanosecond experiments (133). This mechanism was speculated to be due to a MV^{2+}–S^{2-} charge transfer complex. The Zn^{2+}-rich ZnS colloids did not show the presence of the product MV^+. The time to effect MV^{2+} reduction was approximately 20 ps.

Leheney, Rossetti, and Brus (136) studied tetrathiafulvalene (TTF) photoionization as a function of micelle surface charge. Because TTF has negligible solubility in water at neutral pH, it was solubilized near the micelle surface, the environment of which had not been elucidated (137,138). Photoionization of TTF to create TTF^+ enabled its migration from the micelle (occurring over an approximate 10^{-5} second); kinetics of this migration were measured. The C=C

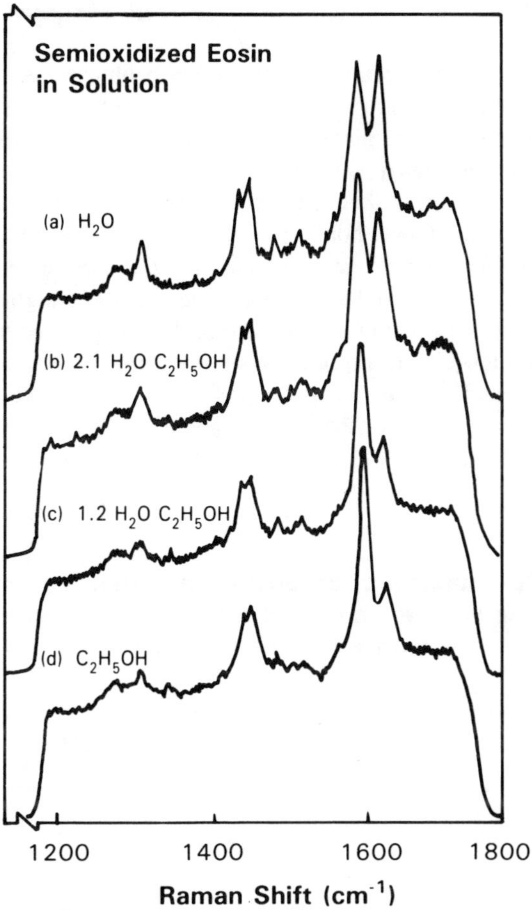

Figure 6.12. Time-resolved Raman spectra of solution-phase semioxidized eosin Y as a function of the solvent at neutral pH. In each case the solution contains 2×10^{-4} *M* eosin and 10^{-2} *M p*-benzoquinone. Excitation at 532 nM (4 mJ/pulse) is followed 50 ns later by a Raman generating probe pulse at 448 nm (0.3 mJ/pulse) (132). Reprinted with permission from *Journal of Physical Chemistry*. Copyright 1984 American Chemical Society.)

stretching mode of TTF^+ was downshifted in the presence of micelles as compared to its value in aqueous solutions. This shift indicated a strong association between a negatively charged micelle and the positive organic ion. Using a positively charged micelle, TTF^+ was found to solvate in the aqueous phase within approximately 5 ns of TTF^+ creation.

Stilbene

The *cis–trans* photoisomerization of stilbene has been used as a model for olefin isomerization (139,140) and has been investigated by use of TR^3 techniques within nanosecond (141) and picosecond (142) time frames. Gustafson, Roberts, and Chernoff (143) also used picosecond TR^3 spectroscopy to study the structure of excited states of *trans*-stilbene, because *cis–trans* isomerization is believed to proceed through the C═C bond configuration of the olefin portion of *trans*-stilbene in its singlet, first excited state, S_1. Band assignments and frequencies of *trans*-stilbene are listed in Table 6.9. Electronic states higher in energy than S_1, labeled S_n, also were elucidated. A significant displacement of electron density in these high electronic states was observed within the olefinic portion of *trans*-stilbene and within the *ortho* and *meta* carbon atoms on the phenyl rings; this description was consistent with previous discussions of charge-resonance excitation scenarios (144). The S_n state was believed to be a pi* state.

Table 6.9 Comparison of ground state vibrational frequencies in *trans*-stilbene with selected excited state frequencies observed using various methods (143).

	EXCITED STATE			
GROUND STATE	ps TRANSIENT RAMAN	ns TRANSIENT RAMAN	LOW TEMPERATURE ABSORPTION	SUPERSONIC JET EXPANSION
228	197	—	209	197
283	284	—	283	291
455	456	—	—	474
619	565	570	592	591
641	621	619	613	620
739	723	—	—	715
764	790	—	—	789
867	845	844	847	846,850,852
999	979	978	978	972
1028	1078	1077	1056	1056
1157	1150	1148	1145	1139–1143
1182	1181	1179	1163	1170
1194	1242	1241	1250	1249
1322,1337	1333	1331	1337	1332
1448	1422	1421	1427	1447
1493	1473	1458	1460	—
1598	1611	—	1599	1610–1628
1639	1567	1534,1566	1549	1545–1555

Quinone Derivatives

Several NR (145–147) and TR^3 studies on quinone derivatives have been published. These derivatives are photosensitizers of photochemical processes, participating in electron transfer and hydrogen atom abstraction reactions. Moore et al. (148) studied anthraquinone-2,6-disulfonate (AQ26DS) in oxygenated aqueous solutions containing nitrite ions. The nanosecond transient Raman spectra showed the presence of the radical anion of AQ26DS with a first-order decay constant of $5.1 \times 10^5\ s^{-1}$.

Tripathi and Schuler (149) investigated the *p*-benzosemiquinone radical ($HOC_6H_4O\cdot$) in pulse radiolysis experiments using hydroquinone under moderately acidic conditions. Strong enhancements in ring stretching and CH bending modes suggested that there was a strong interaction of the OH *p*(pi) electron pair with the pi-electron system of the phenoxyl radical. In strongly acidic solutions, the hydroquinone radical, with spectral characteristics similar to the *p*-benzosemiquinone radical anion, also was observed. This similarity indicated that protonation of the oxygen atoms in the semiquinone system had little effect on its electronic structure.

Qin, Tripathi, and Schuler (150) were able to use TR^3 techniques to examine acid–base equilibria and to follow deprotonation kinetics for 2-hydroxy-*p*-ben-

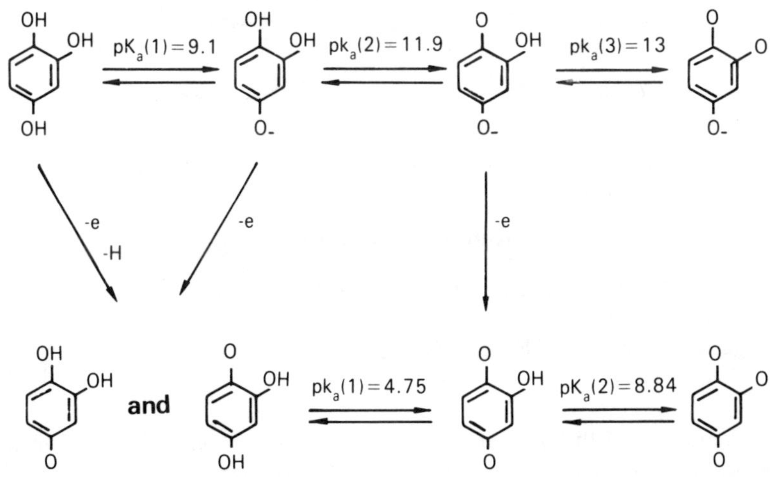

Ia Ib II III

Figure 6.13. Acid–base equilibria of 1,2,4-benzenetriol and between radical intermediates produced on oxidation (150). (Reprinted with permission from *Journal of Physical Chemistry*. Copyright 1987 American Chemical Society.)

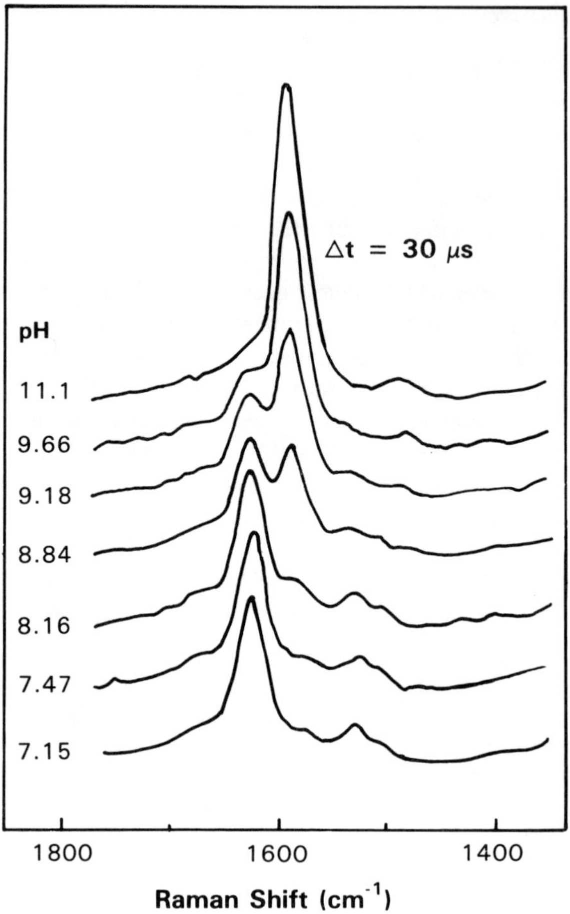

Figure 6.14 The pH dependence of Raman spectra observed 30 μs after excitation at 430 nm (150). (Reprinted with permission from the *Journal of Physical Chemistry*. Copyright 1987 American Chemical Society.)

zosemiquinone and the dianion radicals during oxidation of 1,2,4-benzenetriol. An equilibrium diagram for oxidation of 1,2,4-benzenetriol is shown in Figure 6.13. At high pH, the 1,2,4-benzenetriol was rapidly oxidized by OH or N_3 to form a mixture of neutral 2,4- and 3,4-dihydroxyphenoxyl radicals. For example, Raman data in Figure 6.14 show that 30 μs after pulse irradiation and as the pH of the solution was raised, the intensity of a band at 1625 cm^{-1} decreased while the intensity of a band at 1587 cm^{-1} increased. The band at 1587 cm^{-1} was attributed to a ring stretching mode of the radical dianion, whereas the band at 1625 cm^{-1} was attributed to the monoanion radical. The deprotonation rate of the monoanion upon reaction with base was $(9.6 \pm 1.5) \times 10^9\ M^{-1}s^{-1}$.

Retinal

Hamaguchi et al. (151) studied TR^3 results for 7-, 9-, 11-, and 13-mono-*cis* isomers of retinal in an effort to elucidate the mechanism of *cis–trans* isomerization. Such isomerization is a key step in the visual process and was expected to involve the first excited triplet state, T_1. These results extended previous TR^3 results of all-*trans* retinal (152,153). Triplet state spectra for these 7-, 9-, and 11-*cis* isomers were identical to the ground state spectra of the all-*trans* retinal, indicating that a common relaxed triplet was present for these retinal species. The $C_{13}{=}C_{14}$ double bond is a possible site of isomerization. An all-*trans* T_1 retinal excited state was the most stable isomer of T_1 retinal, whereas the vibrationally relaxed *cis*-triplet species was not believed to be involved in *cis–trans* isomerization.

Other Resonant Studies

Resonance Raman spectroscopy was used to investigate the *trans–cis* photoisomerization of N-methylthioacetamide (NMTAA), which accompanies excitation of the $^1pi^* \leftarrow pi$ transition of the S=C—N chromophore in the *trans* conformer (154,155). Bands of the *cis* and *trans* isomers can be readily distinguished and their relative concentrations quantified by RR techniques. Kato, Hamaguchi, and Tasumi (156) used this information to perform TR^3 experiments on NMTAA in aqueous solutions, spectra of which are displayed in Figure 6.15. Bands at 1461 and 720 cm^{-1} appear upon photoexcitation; the 720 cm^{-1} band previously was assigned to the C=S stretching mode of the *cis* isomer (156). Formation of the *cis* form was described to occur within 5 ns of photoexcitation, and isomerization was believed to be associated with relaxation of the molecule from the $^1pi_1{}^*pi$ to the $^3pi^*n_s$ state by intersystem crossing. However, the importance of the $^1pi^*pi$ state in isomerization could not be discounted.

The RRS and luminescence results for Ni-, Pd-, and Pt tetraphenylporphyrins in noncoordinating and coordinating solvents were reported by Seo, Chung, and Kim (157). Table 6.10 shows band assignments and positions of these porphyrins in piperidine and methylene chloride solvents for these molecules. The use of RRS on this type of complex has enabled investigators to define structural changes as a function of the type of porphyrin and solvent (158). Upon incorporation in strongly ligating solvents, the d^8 Ni^{2+} porphyrin changed from its normal 4-coordinate, low-spin complex to a 6-coordinate, high-spin form. This change in spin state increased the metal–pyrrole N distance and concomitantly caused

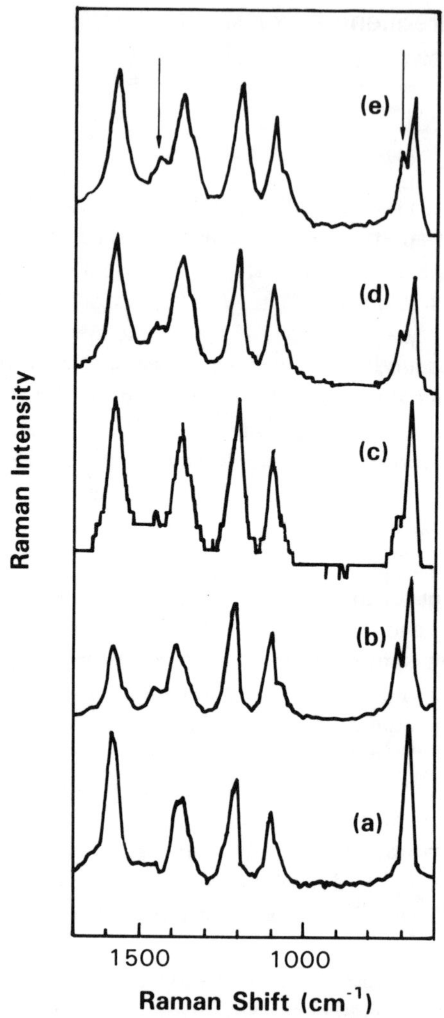

Figure 6.15. Transient resonance Raman spectra of NMTAA in aqueous solutions [1.06 $\times$ 10^{-3} *M* for (a) and (b); 1.1 $\times$ 10^{-2} *M* for (c), (d), and (e)]: (a) obtained with a jet flow; (b) stationary system; (c), (d), and (e) jet flow. The *cis* bands are marked by arrows (156). (Reprinted with permission from the *Journal of Physical Chemistry*. Copyright 1987 American Chemical Society.)

Table 6.10. RR frequencies for Ni, Pd, and PtTPP complexes (157).

		NiTPP		PdTPP		PtTPP	
ASSIGNMENT		CH_2Cl_2	PIP	CH_2Cl_2	PIP	CH_2Cl_2	PIP
A	phenyl	1598	1593	1595	1596	1597	1597
ν_2	C_bC_b	1569	1546	1557	1557	1560	1560
ν_{11}	C_aC_m	1549	1544	—	—	—	—
ν_{20}	C_nC_m	1548	—	—	—	1556	1558
ν_{12}	C_bC_b	1505	1491	—	—	1501	—
ν_4	C_aN	1369	1348	1365	1364	1370	1371
C	$C_m$ph	1233	1231	1233	1233	1233	1233
ν_{15}	δC_bH	1074[a]	—	—	—	1077	1075
E	phenyl	1081	1077	1080	1082	1084	1084
F	phenyl	998	996	1013	—	1031	—
ν_6	C_aC_m	—	989	—	—	1017	1019

[a]From 514.5 nm Ar^+ line excitation.

changes in the vibrational frequencies of the skeletal modes. For Pd and Pt porphyrins, complexation with the piperidine solvent was determined to occur, in which a 5-coordinate, low-spin complex was formed. However, the RR frequencies of these Pd and Pt complexes were very similar to each other as a result of their constant low-spin states in both methylene chloride and piperidine solvents.

REFERENCES

1. Freund, H. J., and E. W. Plummer, *Phys. Rev. B23,* 4859 (1981).
2. Brotman, A., E. Burstein, and J. D. Jiang, *Surf. Sci. 158,* 1 (1985).
3. Morris, D. E., and W. H. Woodruff, in *Spectroscopy of Inorganic-Based Materials,* ed. R. J. H. Clark and R. E. Hester (Wiley New York, 1987), p. 285.
4. Edwards, A. J., W. E. Falconer, J. E. Griffiths, W. A. Sunder, and L. J. Basile, *J. Chem. Soc. Dalton,* 1129 (1974).
5. Lorenzelli, V., *Compt. Rend. 258,* 5836 (1964).
6. Klaboe, P., *J. Am. Chem. Soc. 89,* 3667 (1967).
7. Gratzel, M., and F. P. Rotzinger, *Inorg. Chem. 24,* 2320 (1985).
8. El-Hinnawi, M. A., L. Peter, and B. Meyer, *J. Raman Spectrosc. 16,* 272 (1985).
9. Guidotti, F. B., and A. Giotti, *Bull. Soc. Ital. Biol. Sper. 43,* 1390 (1967).
10. Gabelica, A. Z., *Bull. Acad. R. Belg. (Cl. Sci.) 59,* 1029 (1973).
11. Freedman, A. N., and B. P. Staughan, *Spectrochim. Acta 27A,* 1455 (1971).
12. Hastie, J. W., R. H. Hauge, and J. L. Margrave, *High Temp. Sci. 1,* 76 (1969).
13. Crowder, G. A., G. Gorin, F. H. Kruse, and D. W. Scott, *J. Mol. Spectrosc. 16,* 115 (1965).
14. Caminiti, R., D. Atzei, P. Cucca, A. Anedda, and G. Bongiovanni, *J. Phys. Chem. 90,* 238 (1986).
15. Nelson, D. L., and D. E. Irish, *J. Chem. Phys. 54,* 4479 (1971).
16. Irish, D. E., in *Ionic Interactions,* vol. II, ed. S. Petrucci (Academic Press, New York, 1971).
17. Caminiti, R., and P. Cucca, *Chem. Phys. Lett. 108,* 51 (1984).

18. Irish, D. E., and M. H. Brooker, in *Advances in Infrared and Raman Spectroscopy,* ed. R. J. H. Clark and R. E. Hester (Heyden, London, 1970).
19. Irish, D. E., and T. Jarv, *Appl. Spectrosc. 37,* 50 (1983).
20. Bulmer, J. T., D. E. Irish, and L. Odberg, *Can J. Chem. 53,* 3806 (1975).
21. Sze, Y. K., and D. E. Irish, *J. Sol. Chem. 7,* 417 (1978).
22. Sze, Y. K., and D. E. Irish, *J. Sol. Chem. 8,* 395 (1979).
23. Kanno, H., and J. Hiraishi, *J. Raman Spectrosc. 18,* 157 (1987).
24. Kanno, H., and J. Hiraishi, *Chem. Phys. Lett. 68,* 46 (1979).
25. Kanno, H., and J. Hiraishi, *J. Raman Spectrosc. 12,* 224 (1982).
26. Kabisch, G., I. Bader, H-H. Emons, and K. Pollmer, *J. Mol. Liq. 26,* 139 (1983).
27. Kabisch, G., and K. Pollmer, *J. Mol. Struct. 81,* 35 (1982).
28. Ratcliffe, C. I., and D. E. Irish, *Can. J. Chem. 62,* 1134 (1984).
29. Gillard, R. D., and G. Wilkinson, *J. Chem. Soc.,* 1640 (1964).
30. Goggin, P. L., and T. G. Buick, *Chem. Commun.,* 290 (1967).
31. Basile, L. J., J. R. Ferraro, P. La Bonville, and M. C. Wall, *Coord. Chem. Rev. 11,* 21 (1973).
32. Gonzalez-Vilchez, F., and W. P. Griffith, *J. Chem. Soc. Dalton,* 1416 (1972).
33. Busey, R. H., and O. L. Keller, Jr., *J. Chem. Phys. 41,* 215 (1964).
34. Woodward, L. A., and H. L. Roberts, *Trans. Faraday Soc. 32,* 615 (1956).
35. Dean, K. J., and G. R. Wilkinson, *J. Raman Spectrosc. 14,* 130 (1983).
36. Himeno, S., and M. Hasegawa, *Inorg. Chim. Acta 83,* L5 (1984).
37. Himeno, S., and M. Hasegawa, *Inorg. Chim. Acta 83,* L17 (1984).
38. Himeno, S., M. Hasegawa, and A. Saito, *Inorg. Chim. Acta 101,* L15 (1985).
39. Caldiera, M. M., and J. J. C. T-Dias, *Spectrochem. Acta 40A,* 265 (1984).
40. Kruus, P., A. C. Hayes, and W. A. Adams, *J. Sol. Chem. 14,* 117 (1985).
41. Goggin, P. L., and J. Mink, *J. Chem. Soc. Dalton,* 1479 (1974).
42. Hendra, P. J., *J. Chem. Soc. A,* 1298 (1967).
43. Beattie, I. R., and K. M. Livingston, *J. Chem. Soc. A,* 859 (1969).
44. van Huong, P., and B. Desbat, *Bull. Soc. Chim. Fr.,* 2631 (1972).
45. Bottger, G. L., and C. V. Damsgard, *Spectrochim. Acta 28A,* 1631 (1972).
46. Reisfeld, M. J., *J. Mol. Spectrosc. 29,* 120 (1969).
47. Debeau, M., and M. Krauzman, *Compt. Rend. 264B,* 1724 (1967).
48. Claassen, H. H., G. L. Goodman, J. H. Holloway, and H. Selig, *J. Chem. Phys. 53,* 341 (1970).
49. Woodward, L. A., and J. A. Creighton, *Spectrochim. Acta 17,* 594 (1961).
50. Angell, C. A., and D. M. Gruen, *J. Am. Chem. Soc. 88,* 5192 (1966).
51. Swaddle, T. W., and L. Fabes, *Can. J. Chem. 58,* 1418 (1980).
52. Bell, R. P., and H. C. Longuet-Higgins, *Proc. Roy. Soc. A183,* 357 (1945).
53. Griffith, W. P., and P. J. B. Lesniak, *J. Chem. Soc. A,* 1066 (1969).
54. Brown, R. G., and S. D. Ross, *Spectrochim. Acta 29A,* 169 (1973).
55. Occelli, M. L., and J. M. Stencel, Petroleum Div. Reprints, ACS Natl. Mtg., New Orleans (Sept. 1987).
56. Griffith, W. P., and T. D. Wickins, *J. Chem. Soc. A,* 1087 (1966).
57. Beattie, I. R., and G. A. Ozin, *J. Chem. Soc. A,* 2615 (1969).
58. Chassaing, E., F. Basile, and G. Lordthioir, *J. Appl. Electro-Chem. 11,* 193 (1981).
59. Clark, R. J. H., B. K. Hunter, and D. M. Rippon, *Inorg. Chem. 11,* 56 (1972).
60. van Bronswyk, W., R. J. H. Clark, and L. Maresca, *Inorg. Chem. 8,* 1395 (1969).
61. Wendling, E., and R. Rohmer, *Bull. Soc. Chim. Fr. 1,* 8 (1967).
62. Reichmann, M. G., and A. T. Bell, *Langmiur 3,* 111 (1987).

63. Jones, L. H., R. S. McDowell, M. Goldblatt, and B. I. Swanson, *J. Chem. Phys. 57,* 2050 (1972).
64. Jones, L. H., R. S. McDowell, and M. Goldblatt, *Inorg. Chem. 8,* 2349 (1969).
65. Terzis, A., and T. G. Spiro, *Inorg. Chem. 10,* 643 (1971).
66. Hendra, P. J., and M. M. Qurashi, *J. Chem. Soc. A,* 2963 (1968).
67. Quicksall, C. O., and T. G. Spiro, *Inorg. Chem. 7,* 2365 (1968).
68. Quicksall, C. O., and T. G. Spiro, *Inorg. Chem. 8,* 2363 (1969).
69. Goggin, P. L., and R. J. Goodfellow, *J. Chem. Soc. Dalton,* 2355 (1973).
70. Swanson, B. I., J. J. Rafalko, D. F. Shriver, J. San Filippo, Jr., and T. G. Spiro, *Inorg. Chem. 14,* 1737 (1975).
71. Angel, C. L., F. A. Cotton, B. A. Frenz, and T. R. Webb, *Chem. Commun.,* 399 (1973).
72. Quicksall, C. O., and T. G. Spiro, *Inorg. Chem. 9,* 1045 (1970).
73. Edgell, W. F., J. W. Fisher, G. Asato, and W. M. Risen, *Inorg. Chem. 8,* 1103 (1969).
74. Mays, M. J., and R. N. F. Simpson, *J. Chem. Soc. A,* 1444 (1968).
75. Bennett, M. J., W. A. G. Graham, J. K. Hoyano, and W. L. Hutcheon, *J. Am. Chem. Soc. 94,* 6232 (1972).
76. Ng, J. B., and H. F. Shurvell, *J. Phys. Chem. 91,* 496 (1987).
77. Bulmer, J. T., and H. F. Shurvell, *J. Phys. Chem. 77,* 256 (1973).
78. Malinowski, E. R., R. A. Cox, and U. L. Haldna, *Anal. Chem. 56,* 778 (1984).
79. Hayashi, H., and J. B. Moffatt, *J. Catal. 77,* 473 (1982).
80. Hayashi, H., and J. B. Moffatt, *J. Catal. 81,* 61 (1983).
81. Hayashi, H., and J. B. Moffatt, *J. Catal. 83,* 192 (1983).
82. Misono, M., T. Komaya, H. Sekiguchi, and Y. Yoneda, *Chem. Lett.,* 53 (1982).
83. Keggin, J. F., *Proc. Roy. Soc. London A, 144,* 75 (1934).
84. Lyhamn, L., S. J. Cyvin, B. N. Cyvin, and J. Brunvoll, *Spectrosc. Lett. 9,* 859 (1976).
85. Lyhamn, L., S. J. Cyvin, B. N. Cyvin, and J. Brunvoll, *Spectrosc. Lett. 12,* 101 (1979).
86. Kasprzak, M. S., G. E. Leroi, and S. R. Crouch, *Appl. Spectrosc. 36,* 285 (1982).
87. Ng, K. Y. S., and E. Gulari, *Polyhedron 3,* 1001 (1984).
88. Wilbrandt, R., N. H. Jensen, P. Pagsberg, A. H. Sillesen, and K. B. Hansen, *Nature 276,* 167 (1978).
89. Dallinger, R. F., and W. H. Woodruff, *J. Am. Chem. Soc. 101,* 4391 (1979).
90. Wilbrandt, R., P. Pagsberg, K. B. Hansen, and C. V. Weisberg, *Chem. Phys. Lett. 36,* 76 (1975).
91. Harada, I., Y. Sugawara, H. Matsuura, and T. Shimanouchi, *J. Raman Spectrosc. 4,* 91 (1975).
92. Dallinger, R. F., and W. H. Woodruff, *J. Am. Chem. Soc. 101,* 4391 (1979).
93. Benasson, R., C. Salet, and V. Balzani, *J. Am. Chem. Soc. 98,* 3722 (1976).
94. Wrighton, M. S., L. Pdungsap, and D. L. Morse, *J. Phys. Chem. 79,* 66 (1975).
95. Bradley, P. G., N. Kress, B. A. Hornberger, R. F. Dallinger, and W. H. Woodruff, *J. Am. Chem. Soc. 103,* 7441 (1981).
96. McClanahan, S., T. Hayes, and J. Kincaid, *J. Am. Chem. Soc. 105,* 4486 (1983).
97. Smothers, W. K., and M. S. Wrighton, *J. Am. Chem. Soc. 105,* 1067 (1983).
98. McClanahan, S., T. Hayes, and J. Kincaid, *J. Am. Chem. Soc. 105,* 4486 (1983).
99. Bradley, P. G., N. Kress, B. A. Hornberger, R. F. Dallinger, and W. H. Woodruff, *J. Am. Chem. Soc. 103,* 7441 (1981).
100. Casper, J. V., T. D. Westmoreland, G. H. Allen, P. G. Bradley, T. J. Meyer, and W. H. Woodruff, *J. Am. Chem. Soc. 106,* 3492 (1984).
101. Forster, M., R. B. Girling, and R. E. Hester, *J. Raman Spectrosc. 12,* 36 (1982).
102. Sweetser, P. B., *Anal. Chem. 39,* 979 (1967).

103. Okura, I., and N. K. Thuan, *J. Chem. Soc. Commun.*, 84 (1980).
104. Kalyanasundaram, K., J. Kuvi, and M. Gratzel, *Helv. Chim. Acta 61,* 2720, (1978).
105. Maradpour, A., E. Amouyal, P. Keller, and H. Kagan, *Nouv. J. Chim. 2,* 547 (1978).
106. Kawai, T., K. Tanimura, and T. Sakata, *Chem. Lett.*, 137 (1979).
107. Mabrouk, P. A., and M. S. Wrighton, *Chem. Phys. Lett., 104,* 174 (1984).
108. Ziegler, L. D., and B. Hudson, *J. Chem. Phys. 74,* 982 (1981).
109. Tang, J., and A. C. Albrecht, in *Raman Spectroscopy,* ed. H. A. Syzmanski (Plenum Press, New York, 1970).
110. Korenowski, G. M., L. D. Ziegler, and A. C. Albrecht, *J. Chem. Phys. 68,* 1248 (1978).
111. Ziegler, L. D., and B. S. Hudson, *J. Chem. Phys. 79,* 1134 (1983).
112. Beck, S. M., and L. E. Brus, *J. Chem. Phys. 75,* 4934 (1981).
113. Beck, S. M., and L. E. Brus, *J. Am. Chem. Soc. 103,* 2495 (1981).
114. Grabowska, A., J. Herbich, E. Kirkor-Kaminska, and B. Pakula, *J. Lumin. 11,* 403 (1976).
115. Waluk, J., A. Grabowska, and J. Lipinski, *Chem. Phys. Lett. 70,* 175 (1980).
116. Kobayashi, T., and S. Koshihara, *Chem. Phys. Lett. 104,* 174 (1984).
117. Van Zeyl, P. H. M., C. A. G. O. Varma, and G. Vroege, *Chem. Phys. Lett. 105,* 127 (1984).
118. Tripathi, G. N. R., and M. R. Fisher, *Chem. Phys. Lett. 104,* 297 (1984).
119. Chinsky, L., A. Laigle, W. L. Peticolas, and P.-Y. Turpin, *J. Chem. Phys. 76,* 1 (1982).
120. Blazej, D., and W. L. Peticolas, *J. Chem. Phys. 72,* 3134 (1980).
121. Albrecht, A. C., *J. Chem. Phys. 34,* 1476 (1961).
122. Sufra, G. D., G. Masetti, and G. Zerbi, *J. Raman Spectrosc. 6,* 267 (1977).
123. Inagaki, F., M. Tasumi, and T. Miyazawa, *J. Mol. Spectrosc. 50,* 286 (1974).
124. Blazej, D. C., and W. L. Peticolas, *Proc. Nat. Acad. Sci. USA 74,* 2639 (1977).
125. Warshel, A., and P. Dauber, *J. Chem. Phys. 66,* 5477 (1977).
126. Blazej, D. C., and W. L. Peticolas, *J. Chem. Phys. 72,* 3134 (1980).
127. Asher, S. A., *Anal. Chem. 56,* 720 (1984).
128. Rumelfanger, R., S. A. Asher, and M. B. Perry, *Appl. Spectrosc. 42,* 267 (1988).
129. Beck, S. M., and L. E. Brus, *J. Am. Chem. Soc. 105,* 1106 (1983).
130. Alkaitis, S. A., and M. Gratzel, *J. Am. Chem. Soc. 98,* 3549 (1976).
131. Kratochril, B., and D. Zatko, *Anal. Chem. 40,* 422 (1968).
132. Rosetti, R., and L. E. Brus, *J. Am. Chem. Soc. 106,* 4336 (1984).
133. Rosetti, R., S. M. Beck, and L. E. Brus, *J. Am. Chem. Soc. 106,* 980 (1984).
134. Rosetti, R., Beck, S. M., and L. E. Brus, *J. Am. Chem. Soc. 104,* 7322 (1984).
135. Rosetti, R., and L. E. Brus, *J. Chem. Phys. 90,* 558 (1986).
136. Leheny, A. R., R. Rosetti, and L. E. Brus, *J. Phys. Chem. 89,* 4091 (1985).
137. Kasatani, K., M. Kawasaki, H. Sato, and N. Nakashima, *J. Phys. Chem. 89,* 542 (1985).
138. Turley, D. W., and H. W. Offen, *J. Phys. Chem. 88,* 5542 (1984).
139. Saltiel, J., J. D'Agostino, E. D. Megarity, L. Metts, K. R. Neuberger, M. Wrighton, and O. C. Zafirou, *Org. Photochem. 3,* 1 (1971).
140. Hockstrasser, R. M., *Pure Appl. Chem. 52,* 2683 (1980).
141. Hamaguchi, H., T. Urano, and M. Tasumi, *Chem. Phys. Lett. 106,* 153 (1984).
142. Gustafson, T. L., D. M. Roberts, and D. A. Chernoff, *J. Chem. Phys. 79,* 1559 (1983).
143. Gustafson, T. L., D. M. Roberts, and D. A. Chernoff, *J. Chem. Phys. 81,* 3438 (1984).
144. Yoshihara, K., A. Namiki, M. Sumitani, and N. Nakashima, *J. Chem. Phys. 71,* 2892 (1979).
145. Tripathi, G. N. R., and R. H. Schuler, *J. Chem. Phys. 78,* 2139 (1982).
146. Hester, R. E., and K. P. Williams, *J. Chem. Soc. Faraday Trans. II 78,* 573 (1982).
147. Schuler, R. H., G. N. R. Tripathi, M. F. Prebenda, and D. M. Chipman, *J. Phys. Chem. 87,* 3101 (1983).
148. Moore, J. N., G. H. Atkinson, D. Phillips, P. M. Killough, and R. E. Hester, *Chem. Phys. Lett. 107,* 381 (1984).

149. Tripathi, G. N. R., and R. H. Schuler, *J. Phys. Chem. 91,* 5881 (1987).
150. Qin, L., G. N. R. Tripathi, and R. E. Schuler, *J. Phys. Chem. 91,* 1905 (1987).
151. Hamaguchi, H., H. Okamoto, M. Tasumi, Y. Mukai, and Y. Koyama, *Chem. Phys. Lett. 107,* 355 (1984).
152. Wilbrandt, R., and N.-H. Jensen, *J. Am. Chem. Soc. 103,* 1036 (1981).
153. Atkinson, G. H., J. P. Pallix, T. B. Freeman, D. A. Gilmore, and R. Wilbrandt, *J. Am. Chem. Soc. 103,* 5069 (1981).
154. Sugawara, Y., H. Hamaguchi, I. Harada, and T. Shimanouchi, *Chem. Phys. Lett. 52,* 323 (1977).
155. Harada, I., and M. Tasumi, *Chem. Phys. Lett. 70,* 279 (1980).
156. Kato, C., H. Hamaguchi, and M. Tasumi, *J. Phys. Chem. 89,* 407 (1985).
157. Seo, J.-C., Y.-B. Chung, and D. Kim, *Appl. Spectrosc. 41,* 1199 (1987).
158. Shelnutt, J. A., L. D. Cheung, R. C. C. Chang, N.-T. Yu, and R. H. Felton, *J. Chem. Phys. 66,* 3387 (1979).

7
Future Raman Spectroscopic Applications

The discussion and data present in Chapters 1 through 6 show that Raman spectroscopy has been used successfully to define molecular speciation of potential importance for many catalysts and catalytic reactions. Study of instrumental and experimental factors that enhance the flexibility of Raman spectroscopy—such as, the numerous types of excitation sources that can be used, the application of dispersive or nondispersive delineation of the Raman scattered radiation, the use of simple sampling equipment and sample cell geometries, the acquisition of data under continuous or time-gated conditions and under process realistic conditions, the acquisition of polarization and molecular site-dependent vibrational activity, and so on—has been complemented by the development of sophisticated theoretical treatments. These variables help to overcome the normally low scattering cross section of Raman radiation, but they sometimes introduce unwanted complexity in the Raman spectra. The complex band structure and the multiplicity of bands that are found can supplement the experimental information if they can be interpreted. The data base for Raman spectroscopic applications to catalysis is relatively small (as compared to, for example, infrared spectroscopy)—in fact, just in its initial stage of development for resonance Raman and surface enhanced Raman spectroscopy; so the application of Raman spectroscopy to catalytic problems seems very promising.

A particular advantage of Raman spectroscopy is that it monitors only specific molecular sites in a material. This specificity is unlike that of electron or ion spectroscopic techniques, where any particular element is detected independent of its geometric or electronic configuration. Molecular species having distorted symmetry due to interactions with a surface or incorporation into a lattice network produce vibrational spectra at variance with nondistorted species. Observation of such species has been difficult because of spectral obstacles such as fluorescence or low Raman intensities. The former problem needs to be more thoroughly understood, but it can be overcome by the use of near-infrared laser excitation and Fourier transform Raman spectroscopy (FTRS). Commercial FTRS instrumentation now being marketed reportedly is as sensitive as normal Raman spectroscopy while using visible excitation. The state of development of FTRS could be compared to the development of Fourier transform infrared spectroscopy

(FTIR) approximately a decade ago; currently, most new infrared spectrometers that are purchased are of the FTIR type.

The problem of low sensitivity in Raman spectroscopy can be overcome by the use of tunable laser sources and resonance effects. One such truly broadly tunable laser source is the free electron laser; a national center currently under construction at Vanderbilt University will have a free electron laser for use by interested scientists.

As discussed in Chapter 4, normal Raman spectroscopy has been applied primarily to the oxidized form of catalysts because the normal black color of activated catalysts absorbs the incident and scattering radiation. Hence, a tremendous amount of vibrational spectroscopic research has yet to be performed on supported catalysts in their activated state (i.e. sulfided, carbided, reduced, etc.). Also, adsorption on bimetallic or alloy catalysts and detection of the adsorbed species has not been addressed adequately by Raman techniques. Using FTRS and/or free electron laser techniques, Raman spectroscopy might produce new and exciting results on the structural/chemical forms of these activated catalysts and on adsorbed species.

Catalyst preparation and the species present during catalyst synthesis have not been examined thoroughly with the new and developing Raman techniques. For example, the observation by Raman spectroscopy of the chemistry of zeolite synthesis in solutions is in its infancy. The delineation of cation–anion interactions in aqueous and nonaqueous solutions is another area that can be affected by Raman spectroscopy.

As shown in Chapter 4, catalyst deactivation and the structural and/or chemical species associated with such deactivation can be delineated by the use of Raman spectroscopy. However, very little research on catalyst deactivation has utilized Raman spectroscopy, perhaps because of the normal black color of such catalysts. Carbon deposits (size and functionality), metals contamination, zeolite crystallinity and/or transformations, and identification of adsorbed poisons could be investigated with the proper choice of Raman instrumentation. Fluid catalytic cracking and partial oxidation catalysts should be particularly amenable to such directed research.

The summary of surface enhanced Raman (SER) and resonance Raman (RR) data in Chapters 5 and 6 also directs one to potentially fruitful Raman spectroscopic applications. In particular, catalysts containing finely divided silver, copper, or gold, and idealized catalysts in which these metals are coated with a thin layer of other metals, are amenable to examination by Raman techniques. It is possible that thin layers of noble metals or alloys on these SER-activating metals will provide sufficient spectral enhancement for researchers to observe Raman active modes of adsorbed species.

Index